LONDON MATHEMATICAL SOCIETY LECTURE NOTE SERIES

Managing Editor: Professor J.W.S. Cassels, Department of Pure Mathematics and Mathematical Statistics, University of Cambridge, 16 Mill Lane, Cambridge CB2 1SB, England

The books in the series listed below are available from booksellers, or, in case of difficulty, from Cambridge University Press.

4 Algebraic topology, J.F. ADAMS
17 Differential germs and catastrophes, Th. BROCKER & L. LANDER
27 Skew field constructions, P.M. COHN
34 Representation theory of Lie groups, M.F. ATIYAH *et al*
36 Homological group theory, C.T.C. WALL (ed)
39 Affine sets and affine groups, D.G. NORTHCOTT
40 Introduction to H_p spaces, P.J. KOOSIS
42 Topics in the theory of group presentations, D.L. JOHNSON
43 Graphs, codes and designs, P.J. CAMERON & J.H. VAN LINT
45 Recursion theory: its generalisations and applications, F.R. DRAKE & S.S. WAINER (eds)
46 p-adic analysis: a short course on recent work, N. KOBLITZ
49 Finite geometries and designs, P. CAMERON, J.W.P. HIRSCHFELD & D.R. HUGHES (eds)
50 Commutator calculus and groups of homotopy classes, H.J. BAUES
51 Synthetic differential geometry, A. KOCK
54 Markov processes and related problems of analysis, E.B. DYNKIN
57 Techniques of geometric topology, R.A. FENN
58 Singularities of smooth functions and maps, J.A. MARTINET
59 Applicable differential geometry, M. CRAMPIN & F.A.E. PIRANI
60 Integrable systems, S.P. NOVIKOV *et al*
62 Economics for mathematicians, J.W.S. CASSELS
65 Several complex variables and complex manifolds I, M.J. FIELD
66 Several complex variables and complex manifolds II, M.J. FIELD
68 Complex algebraic surfaces, A. BEAUVILLE
69 Representation theory, I.M. GELFAND *et al*
74 Symmetric designs: an algebraic approach, E.S. LANDER
76 Spectral theory of linear differential operators and comparison algebras, H.O. CORDES
77 Isolated singular points on complete intersections, E.J.N. LOOIJENGA
78 A primer on Riemann surfaces, A.F. BEARDON
79 Probability, statistics and analysis, J.F.C. KINGMAN & G.E.H. REUTER (eds)
80 Introduction to the representation theory of compact and locally compact groups, A. ROBERT
81 Skew fields, P.K. DRAXL
82 Surveys in combinatorics, E.K. LLOYD (ed)
83 Homogeneous structures on Riemannian manifolds, F. TRICERRI & L. VANHECKE
84 Finite group algebras and their modules, P. LANDROCK
85 Solitons, P.G. DRAZIN
86 Topological topics, I.M. JAMES (ed)
87 Surveys in set theory, A.R.D. MATHIAS (ed)
88 FPF ring theory, C. FAITH & S. PAGE
89 An F-space sampler, N.J. KALTON, N.T. PECK & J.W. ROBERTS
90 Polytopes and symmetry, S.A. ROBERTSON
91 Classgroups of group rings, M.J. TAYLOR
92 Representation of rings over skew fields, A.H. SCHOFIELD
93 Aspects of topology, I.M. JAMES & E.H. KRONHEIME
94 Representations of general linear groups, G.D. JAMES
95 Low-dimensional topology 1982, R.A. FENN (ed)

96 Diophantine equations over function fields, R.C. MASON
97 Varieties of constructive mathematics, D.S. BRIDGES & F. RICHMAN
98 Localization in Noetherian rings, A.V. JATEGAONKAR
99 Methods of differential geometry in algebraic topology, M. KAROUBI & C. LERUSTE
100 Stopping time techniques for analysts and probabilists, L. EGGHE
101 Groups and geometry, ROGER C. LYNDON
103 Surveys in combinatorics 1985, I. ANDERSON (ed)
104 Elliptic structures on 3-manifolds, C.B. THOMAS
105 A local spectral theory for closed operators, I. ERDELYI & WANG SHENGWANG
106 Syzygies, E.G. EVANS & P. GRIFFITH
107 Compactification of Siegel moduli schemes, C-L. CHAI
108 Some topics in graph theory, H.P. YAP
109 Diophantine Analysis, J. LOXTON & A. VAN DER POORTEN (eds)
110 An introduction to surreal numbers, H. GONSHOR
111 Analytical and geometric aspects of hyperbolic space, D.B.A.EPSTEIN (ed)
112 Low-dimensional topology and Kleinian groups, D.B.A. EPSTEIN (ed)
113 Lectures on the asymptotic theory of ideals, D. REES
114 Lectures on Bochner-Riesz means, K.M. DAVIS & Y-C. CHANG
115 An introduction to independence for analysts, H.G. DALES & W.H. WOODIN
116 Representations of algebras, P.J. WEBB (ed)
117 Homotopy theory, E. REES & J.D.S. JONES (eds)
118 Skew linear groups, M. SHIRVANI & B. WEHRFRITZ
119 Triangulated categories in the representation theory of finite-dimensional algebras, D. HAPPEL
120 Lectures on Fermat varieties, T. SHIODA
121 Proceedings of *Groups - St Andrews 1985*, E. ROBERTSON & C. CAMPBELL (eds)
122 Non-classical continuum mechanics, R.J. KNOPS & A.A. LACEY (eds)
123 Surveys in combinatorics 1987, C. WHITEHEAD (ed)
124 Lie groupoids and Lie algebroids in differential geometry, K. MACKENZIE
125 Commutator theory for congruence modular varieties, R. FREESE & R. MCKENZIE
126 Van der Corput's method for exponential sums, S.W. GRAHAM & G. KOLESNIK
127 New directions in dynamical systems, T.J. BEDFORD & J.W. SWIFT (eds)
128 Descriptive set theory and the structure of sets of uniqueness, A.S. KECHRIS & A. LOUVEAU
129 The subgroup structure of the finite classical groups, P.B. KLEIDMAN & M.W.LIEBECK
130 Model theory and modules, M. PREST
131 Algebraic, extremal & metric combinatorics, M-M. DEZA, P. FRANKL & I.G. ROSENBERG (eds)
132 Whitehead groups of finite groups, ROBERT OLIVER
133 Linear algebraic monoids, MOHAN S. PUTCHA
134 Number thoery and dynamical systems, M. DODSON & J. VICKERS (eds)
135 Operator algebras and applications, 1, D. EVANS & M. TAKESAKI (eds)
136 Operator algebras and applications, 2, D. EVANS & M. TAKESAKI (eds)

London Mathematical Society Lecture Note Series. 136

Operator Algebras and Applications

Volume II: Mathematical Physics and Subfactors

Edited by

DAVID E. EVANS
University of Wales, Swansea

MASAMICHI TAKESAKI
University of California, Los Angeles

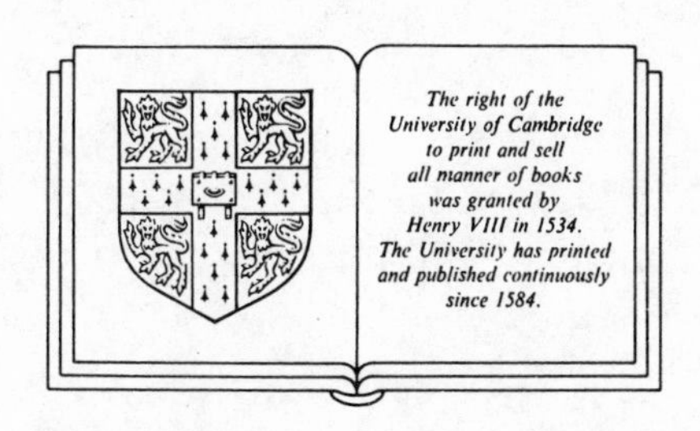

CAMBRIDGE UNIVERSITY PRESS
Cambridge
New York New Rochelle Melbourne Sydney

Published by the Press Syndicate of the University of Cambridge
The Pitt Building, Trumpington Street, Cambridge CB2 1RP
32 East 57th Street, New York, NY 10022, USA
10, Stamford Road, Oakleigh, Melbourne 3166, Australia

First published 1988

Printed in Great Britain at the University Press, Cambridge

Library of Congress cataloging in publication data available

British Library cataloguing in publication data available

ISBN 0 521 36844 X

Preface

A symposium was organised by D.E. Evans at the Mathematics Institute, University of Warwick, between 1st October 1986 and 29th October 1987, with support from the Science and Engineering Research Council, on operator algebras and applications and connections with topology and geometry (K–theory, index theory, foliations, differentiable structures, braids, links) with mathematical physics (statistical mechanics and quantum field theory) and topological dynamics.

As part of that programme, a UK-US Joint Seminar on Operator Algebras was held during 20-25 July 1987 at Warwick, with support from SERC and NSF and organised by D.E. Evans and M. Takesaki. These two volumes contains papers, both research and expository articles, from members of that special week, together with some articles by D.B. Abraham, A.L. Carey, and A. Wassermann on work discussed earlier in the year.

We would like to take this opportunity to thank SERC and NSF for their support, and the participants, speakers and authors for their contributions.

D.E. Evans
Department of Mathematics & Computer Science
University College of Swansea
Singleton Park
Swansea SA2 8PP
Wales, U.K.

M. Takesaki
Department of Mathematics
University of California at Los Angeles
405 Hilgard Avenue
California 90024
U.S.A.

Contents

Volume II: Mathematical Physics and Subfactors.

Volume I: Structure Theory; K-Theory, Geometry and Topology.

UK-US Joint Seminar on Operator Algebras
Lecture list

UK-US Joint Seminar on Operator Algebras
Lectures

SOME RECENT RESULTS FOR THE PLANAR ISING MODEL

D.B. Abraham
Department of Mathematics, University of Arizona,
Tucson, AZ 85721
(on leave from Department of Theoretical Chemistry,
South Parks Road, University of Oxford, Oxford)

Introduction

The planar Ising model has become one of the most important statistical mechanical systems for the study of phase transitions and critical phenomena. Although there are many rigorous results, such as correlation inequalities, Peierls argument and the Yang-Lee circle theorem to name but three [reviewed by Griffiths, 1971], which are dimension-independent in their validity and which lead to results of considerable interest, only the planar model to date benefits from the added insights which stem ultimately from Onsager's tour de force [Onsager, 1944]. It is not the purpose of this article to enter into a general review - for this the reader is referred elsewhere [Gallavotti, 1972] but rather to discuss two more mathematical aspects of the development of Onsager's solution. The first item is the Yang-Baxter system of equations for the planar Ising model in zero field with transfer in the (1,1) direction. This work shows that the Clifford-algebraic structure of the exact solution is a natural consequence of the star-triangle equations. The second item is a Fredholm system which turns out to be of crucial importance in understanding surface and interface problems, as well as the pair correlation function.

Consider a planar lattice drawn on $\mathbb{Z}^2$ and denoted by $\Lambda(N,M) = \{(x,y): -N \le x \le N, -M \le x \le M\}$. At each lattice site (x,y) there is a spin variable $\sigma(x,y) = \pm 1$. The specification of all such spin variables on $\Lambda(N,M)$ is called a spin configuration denoted $\{\sigma\}_{N,M}$. It has an energy

$$E_\Lambda(\{\sigma\}_{N,M}) = -\sum_{|\underline{i}-\underline{j}|=1} J(|\underline{i}-\underline{j}|)\,\sigma(\underline{i})\,\sigma(\underline{j}) \quad - H\,\Sigma\,\sigma(\underline{i}) \tag{1.1}$$

The summations are over points of $\Lambda(N,M)$. We have $J(|\underline{k}|) > 0$ for $|\underline{k}| = 1$; this is the ferromagnetic case. It will turn out to be useful to introduce boundary fields $H(\underline{i})$ which supplement the constant field H. They produce a boundary energy

$$E_B(\{\sigma\}_{N,M}) = -\sum_{\underline{i}\in\partial\Lambda} H(\underline{i})\,\sigma(\underline{i}) \tag{1.2}$$

The canonical configurational probability is

$$p_\Lambda(\{\sigma\}_{N,M}) = Z_\Lambda^{-1} \exp - \beta\, E_\Lambda(\{\sigma\}_{N,M}) \tag{1.3}$$

where β is the inverse temperature in units of the Boltzmann constant and Z_Λ normalises (1.3).

Consider the one-particle correlation function $\langle\sigma(0,0)\rangle(B,N,M)$ where $\langle . \rangle$ is expectation with respect to the measure (1.3) and B denotes that boundary fields have been included. First, let us consider the case $H = 0$. At low enough temperatures, the Peierls argument, which is reviewed by Griffiths [1971] shows that

$$\lim_{N,M\to\infty} \langle\sigma(0,0)\rangle(+,N,M) > 0 \tag{1.4}$$

where + means that all spins on the boundary $\partial\Lambda$ are fixed to be +1 . Since

$$\lim_{N,M\to\infty} \langle\sigma(0,0)\rangle (+,N,M) = - \lim_{N,M\to\infty} \langle\sigma(0,0)(-,N,M) \qquad (1.5)$$

by an obvious symmetry (recall that $H = 0$), the system is unstable with respect to its boundary specification; this is one aspect of phase transition. Onsager´s approach may be developed to give the left hand side of (1.4), which is called the spontaneous magnetisation, for all values of β, in particular those near the critical value βc [Abraham and Martin-Löf 1973].

Another example is as follows: no matter how the boundary conditions B are chosen, we cannot have translationally-variant correlation functions in the planar Ising model in the limit where the boundaries recede to infinity (with mild technical restrictions). This result is due to Aizenman [1979,1980] and to Higuchi [1979] and is true for all β . It can be shown that if the spins are +1 round the boundary above the line $i_2 = 0$ but -1 below, then the magnetisation satisfies the limiting law [Abraham and Reed, 1976]

$$\lim_{N\to\infty} \lim_{M\to\infty} \langle\sigma(\delta N,\alpha M^{1/2})\rangle = m^*\operatorname{sgn}\alpha \quad \Phi\left(\frac{b|\alpha|}{\sqrt{1-\delta^2}}\right) \qquad (1.6)$$

for all $|\delta| < 1$, where

$$\Phi(x) = \frac{2}{\sqrt{\pi}} \int_0^x e^{-u^2} du \qquad (1.7)$$

and

$$b = (\sinh 2(K-K^*))^{1/2} \qquad (1.8)$$

This is less general, but contains more information pertinent to physics, in particular on how the traditional van der Waals-Maxwell picture breaks down. In this picture, the interface between the two phases which coexist (for H=0) for $\beta > \beta_c$ is localised in laboratory-fixed axes.

The n-point correlation functions for local observables $0_1 \ldots 0_n$ which just act on single rows of spins are written as

$$<0_1(\underline{r}_1) \ldots 0_n(\underline{r}_n)> (B,N.M)$$

$$= \frac{<\phi_1|V^{x_1}0_1(y_1)V^{x_2-x_1}0_2(y_2) \ldots V^{(x_n-x_{n-1})}0_n(y_n)V^{N-x_n}|\phi_2>}{<\phi_1|V^N|\phi_2>} \quad (1.9)$$

In the above, V is the transfer matrix [Schultz et al 1964; Abraham 1978a)] which has been symmetrised for spectral convenience. Also, the independence of V on x means that the boundary conditions B at the points $y = \pm m$ are translationally invariant. In what follows, we shall take cyclic boundary conditions although this is not necessary [Abraham and Martin-Löf 1973], but the symmetry is very convenient, particularly in sections 3 and 4 below. The vectors $|\phi_j>$ describe the faces of the cylinder; the scalar products produce the desired weighted sums over end-states provided the $|\phi_j>$ are chosen correctly [see, for instance Abraham and Reed, 1976 for the derivation of (1.6)]. Since N and M are finite, (1.9) makes perfect sense. It can be analysed by knowing the spectrum of V and matrix elements of the local operators $0_j(y)$ in the basis of eigenvector of V. The first follows from Onsager [1944] and simplifications due to Kaufman [1949], Onsager and Kaufman

[1949] and finally Schultz, Mattis and Lieb [1964]. Since then, the matrix elements have been found [Abraham, 1978 a,b,c]. The only remaining problem is the evaluation of scalar products from the eigenvector basis to the $|\phi_j\rangle$. Sometimes the $|\phi_j\rangle$ are rather atypical of the equilibrium state, so that great care is needed with the $M \to \infty$ limit.

2. Monodromy and Yang-Baxter equations

Let us consider the transfer matrix T which works in the (1.1) direction:

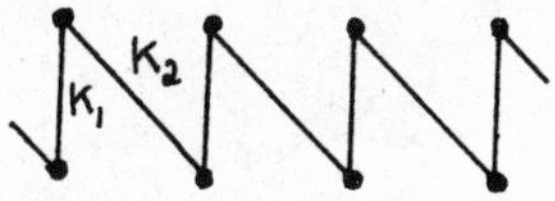

The couplings in units of β are denoted K_j and are arranged as shown.

Each diagonal bond between rows is decorated with an auxiliary spin labelled τ using the equation

$$e^{K_2 \sigma_1 \sigma_2} = \rho \sum_{\tau} e^{(L_1 \sigma_1 + L_2 \sigma_2)\tau} \tag{2.1}$$

Thus the transfer matrix is

$$T = \mathrm{Tr}\, \mathbb{M} \tag{2.2}$$

where the trace is taken over auxiliary spins and the monodromy matrix is

$$\mathbb{M} = \prod_{1}^{n} L_j \tag{2.3}$$

where the product is a matrix one over auxiliary variables which is ordered with increasing j to the right and

$$L_j(\tau_j, \tau_{j+1}) = \exp(K_1 \sigma_j \sigma_j' + L_1 \sigma_j \tau_j + L_2 \sigma_j' \tau_{j+1}) \tag{2.4}$$

Thus we may think of the L_j as 2 x 2 matrices whose entries themselves are matrix-valued in the variables which describe the original rows of spins; these spins are normally referred to as quantum ones. Each element of L_j is now taken as a matrix representative of an operator in the two-dimensional complex vector space $\mathbb{C}^2$, we deal with the product in (2.3) in a tensor product space

$$\mathfrak{h}_M = \bigotimes_1^M \mathbb{C}^2 \tag{2.5}$$

using the Pauli spin operators

$$\sigma_j^\alpha = \bigotimes_1^{j-1} 1 \otimes \sigma^\alpha \bigotimes_{j+1}^M 1 \tag{2.6}$$

where σ^α are the usual Pauli matrices.

After some algebra [Abraham and Davies, 1988], (2.3) and (2.4) become

$$\mathbb{M} = (2 \sinh 2K_1)^{M/2} \begin{pmatrix} \cosh K_2 \\ i \sinh K_2 \sigma_1^x \end{pmatrix} \tilde{T}(1, -\sigma_M^x) \tag{2.7}$$

where

$$\tilde{T} = \prod_1^{2M-1} U_j \tag{2.8}$$

with

$$U_{2j-1} = \exp - K_1^* \sigma_j^z \tag{2.9}$$

$$U_{2j} = \exp K_2 \sigma_j^x \sigma_{j+1}^x \tag{2.10}$$

and the matrices L_j are recovered in the representation with σ_j^x diagonal. The operator (2.7) can be analysed using the Jordan-Wigner transformation

$$f_j^\dagger = P_{j-1}(\sigma_j^x + i\sigma_j^y)/2 \tag{2.11}$$

with

$$P_0 = 1, \quad P_k = \prod_1^k (-\sigma_j^z) \tag{2.12}$$

The associated spinors

$$\Gamma_{2j-1} = f_j^\dagger + f_j, \quad \Gamma_{2j} = -i(f_j^\dagger - f_j) \tag{2.13}$$

generate a Clifford algebra through the anticommutation relations

$$[\Gamma_j, \Gamma_k]_+ = 2\delta_{jk} \tag{2.14}$$

which follow because

$$[f_j, f_k]_+ = 0, \quad [f_j, f_k^{\dagger}]_+ = \delta_{jk} \tag{2.15}$$

as is easily checked.

The factors U_j become spinor rotations

$$U_{2j-1} = \exp i K_1^* \Gamma_{2j-1} \Gamma_{2j} \tag{2.16}$$

$$U_{2j} = \exp i K_2 \Gamma_{2j} \Gamma_{2j+1} \tag{2.17}$$

and

$$\mathbb{M} = (2 \sinh 2K_1)^{M/2} \begin{pmatrix} \cosh K_2 \\ - \sinh K_2 \Gamma_1 \end{pmatrix} \tilde{T}(1, - i\Gamma_{2M} P_M) \tag{2.18}$$

Thus we have

$$T = (2 \sinh 2K_1)^{M/2} \{\cosh K_2 \tilde{T} + i \sinh K_2 \Gamma_1 \tilde{T} \Gamma_{2M} P_M\} \tag{2.19}$$

Returning to (2.12), it is clear that

$$[T, P_M]_- = 0 \tag{2.20}$$

which can be thought of as a statement of rotational invariance through an angle π about the Z-axis, or equally well, of conservation of parity of fermion number. Introducing the projectors

$$Q_{\pm} = (1 \pm P_M)/2 \tag{2.21}$$

onto the sub-spaces $\mathfrak{h}_M^{\pm}$ with even/odd numbers of fermions gives

$$T = T_+ Q_+ + T_- Q_- \tag{2.22}$$

with

$$T_{\pm} = (2 \sinh 2K_1)^{M/2} \{\cosh K_2 \tilde{T} + i \sinh K_2 \Gamma_1 \tilde{T} \Gamma_{2M}\} \tag{2.23}$$

which can be brought to diagonal form using the Schultz-Mattis-Lieb [1964] or Kaufman-Onsager [1949] procedures. The only difference from the usual case is that

(2.23) is a sum of spinor rotations, up to a factor. After some algebra [Abraham and Davies 1988], the result is

$$T_{\pm} = \prod_{\omega\in\Omega_{M(\mp)}} (c_1c_2 + |1 - s_1s_2e^{i\omega}|)^{1/2} \exp - \sum_{\omega\in\Omega_{M(\mp)}} \gamma(\omega)G^{+}(\omega)G(\omega) \tag{2.24}$$

where

$$\Omega_M(\pm) = \{\omega \in (-\pi,\pi],\ \exp iM\omega = \pm 1\} \tag{2.25}$$

and

$$G^{+}(\omega) = U_{\pm}F(\omega)^{+}U_{\pm}^{+}, \quad \omega \in \Omega_M(\pm) \tag{2.26}$$

with

$$F^{+}(\omega) = M^{-1/2}\sum_{1}^{M} e^{i\omega j} f_j^{+} \tag{2.27}$$

and

$$U_{\pm} = \begin{Bmatrix} 1 \\ U_0 \end{Bmatrix} \exp i \sum_{\omega\in(0,\pi)\cap\Omega_M(\mp)} \Theta(\omega)(F^{+}(-\omega)F^{+}(\omega) + F(\omega)F(-\omega)) \tag{2.28}$$

with

$$U_0 = \exp i\,\Theta(0)(F(0)^{+} + F(0)) \tag{2.29}$$

The transformation angle is given by

$$e^{2i\Theta(\omega)} = (1 - s_1s_2e^{i\omega})/|1 - s_1s_2e^{i\omega}| \tag{2.30}$$

and the eigenvalues are constructed in terms of

$$e^{-\gamma(\omega)} = (c_1c_2 - |1 - s_1s_2e^{i\omega}|)/(s_1 + s_2e^{i\omega}) \tag{2.31}$$

(note $|\exp - (\omega)| < 1$ by choice).

In (2.30) and (2.31) we use the convenient notation

$$c_j = \cosh 2K_j \quad s_j = \sinh 2K_j \quad j = 1,2 \tag{2.32}$$

It is interesting to note directly from (2.30) that transfer matrices with the same value of $k = 1 / s_1s_2$ form a commuting family; this was known to Onsager [1944, 1971] and to Stephen and Mittag [1972]. The thermodynamic critical point is given by $k = 1$; the nature of the spectrum clearly depends crucially on the sign of $(k - 1)$: if $k > 1$, $\exp 2i\theta(0) = 1$ so $U_0 = 1$ whereas if $k < 1$ (low-temperature) $\exp 2i\theta(0) = -1$ so that $U_0 = i(F_0^+ + F_0)$. Reference to (2.21) to (2.23) allows one to construct the spectrum. Notice that for $k < 1$ there are no states with an odd number of $G^+(\omega)$-created fermions.

The behaviour of the transformation angle mentioned above gives the function

$$\Theta(e^{i\omega}) = \exp(-2i\theta(\omega)) \tag{2.33}$$

an interesting behaviour; the winding number

$$I(\Theta) = \frac{1}{2\pi} \Delta_{|z|=1}(\arg \Theta(z)) \tag{2.34}$$

has the value zero for $k > 1$ but -1 if $k < 1$. This turns out to have important consequences for the matrix-element theory described in the next section.

Let us return meanwhile to (2.7); the reader familiar with the quantum inverse scattering method (QISM : the reader is referred for reviews to Takhtadzhan and Fadeev [1979] and to Thacker [1981] will find the matrix structure in the auxiliary variables suggestive. In the QISM, the off-diagonal elements of $\mathbb{M}$ generate the eigenvector of T by behaving somewhat like ladder operators acting on a vacuum. It turns out that if k is held fixed, but the ratio $-s_2/s_1$ is given a unimodular complex value $e^{i\omega}$, then

$$G^+(\omega) = \alpha\, \mathbb{M}_{21} + \beta\, \mathbb{M}_{12} P_M \tag{2.35}$$

where α and β are given by tedious algebra. It is easy to confirm that the G-operators have c-number-valued anticommutators which only become fermionic for a suitable choice of the ω. Thus we expect M_{12} and M_{21} to have an associated algebra; in the QISM it is natural to inquire whether such a structure can be obtained directly from Yang-Baxter equations [Baxter, 1982]. These are of the form

$$L_j \otimes L_j' = L_j' \otimes L_j \tag{2.36}$$

with R non-singular. Here the direct product is in auxiliary space, the order of quantum operators is respected and R is a 4 x 4 matrix which must be found. Its existence is assured [Gutkin, 1986] since we have a commuting family of transfer matrices; all we need to construct it is the star-triangle relation discovered by Onsager [Onsager 1944, Baxter 1982]. This work is discussed in much more detail elsewhere [Abraham and Davies, 1988]. The point we wish to make here is that (2.36) does not establish a serious alternative calculational procedure to the ones used, rather it explains in a simple way why the Jordan-Wigner transformation is encountered since the algebra of the $\mathbb{M}_{ij}$ can be got.

Ultimately the techniques of the next section might be extended to get the n-point functions of the vertex models of Lieb [for a review, see Lieb and Wu 1971] and of Baxter [Baxter 1982, and references therein].

3. Matrix elements

Let us now turn to the problem of the matrix elements $\langle \phi_j | 0 | \phi_k \rangle$ which will be required in the spectral decomposition of (1.9) if $[0,P_M]_- = 0$ then matrix elements are only non-zero if $\phi_j,\phi_k \in \mathfrak{h}_M^+$ or $\phi_j,\phi_k \in \mathfrak{h}_M^-$. Such an 0 will be a polynomial of even degree in the Fermi operators and the matrix elements can be obtained in principle at least, from the Wick theorem. A typical example here is the energy density.

The other case, when $[0,P_M]_+ = 0$ is a great deal more difficult, but has been solved [Abraham, 1978 a,b,c]. A typical case here for 0 is the local magnetisation σ_j^x. For ease of notation we shall label $\omega \in \Omega_M(-)$ (resp. $\omega \in \Omega_M(+)$) by β (resp. α). Typically, we want

$$<\Phi_-|G(\alpha_1) \ldots G(\alpha_n)\, \sigma_j^x\, G^\dagger(-\beta_{n+1}) \ldots G^\dagger(-\beta_n)|\Phi_+>$$

where the vector on the left (resp. right) is in $\mathfrak{h}_M^-$ (resp. $\mathfrak{h}_M^+$). Such vectors are also eigenvectors of the translation τ defined by

$$\tau\, \sigma_j^x\, \tau^{-1} = \sigma_{j-1}^x \qquad M \geq j \geq 2$$

$$\tau\, \sigma_1^x\, \tau^{-1} = \sigma_M^x \tag{3.1}$$

as explained in Abraham [1978a]. Thus the basic objects are

$$<\Phi_-|G(\alpha_1) \ldots G(\alpha_n)\, \sigma_1^x\, G^\dagger(-\beta_n) \ldots G^\dagger(-\beta_{n+m})|\Phi_+>$$

The advantage now is that $\sigma_1^x = f_1^+ + f_1$ is a linear form in the spinors. Returning to (2.27), there is a linear dependence

$$F^\dagger(\alpha) = \frac{2}{M} \sum_\beta \frac{1}{e^{i(\beta-\alpha)}-1} F^\dagger(\beta) \tag{3.2}$$

and from (2.26) and (2.28) we have

$$G(\beta) = \frac{2}{M} \quad \frac{1}{e^{i(\alpha-\beta)}-1} \{\cos(\theta(\beta)-\theta(\alpha))G(\alpha) + i \sin(\theta(\beta)-\theta(\alpha))G^\dagger \tag{3.3}$$

Thus the key matrix elements are

$<\Phi_-|G(\alpha_n) \ldots G(\alpha_1)|\Phi_+>$ from which the others follow by using (3.3) and the vacuum properties of $|\Phi_\pm>$. In this section we shall give some improvements over the original exposition which occured to the author in the intervening time, mostly stimulated by the desire to understand various surface properties of the planar Ising model.

The first area is the taking of the thermodynamic limit. We define

$$F_M((e^{i\alpha})_n) = M^{n/2} \exp i \sum_{j=1}^{n} (\alpha_j + \theta(\alpha_j)) < \Phi_- | G(\alpha_n) \ \dots \ G(\alpha_1) | \Phi_+) \tag{3.4}$$

It has been shown that [Abraham 1978a]

$$\frac{2}{M} \sum_{z_1 \in S_M^+} F_M((z)_n) \frac{1}{z_1/t-1} \left(1 + \frac{\Theta(z_1)}{\Theta(t)} \right)$$

$$= \sum_{2}^{n} (-1)^j \ h(t, z_j) \ F_M(\Delta_{1j}(z)_n) \tag{3.5}$$

where

$$h(t,z) = \frac{2zt}{zt-1} \left(\frac{1}{\Theta(z)\ \Theta(t)} - 1 \right) \tag{3.6}$$

and $H(z)$ is given by (2.33). The summation is over $\{z : z^M = -1\}$. We use the notation

$(z)_n = z_1, \dots z_n$ and

$$\Delta_j(z)_n = z_1 \dots z_{j-1} z_{j+1} \dots z_n \quad \text{for} \quad 1 \le j \le n \tag{3.7}$$

with

$$\Delta_I = \prod_{j \in I} \Delta_j \tag{3.8}$$

for any index set I . If $n = 1$, the r.h.s. of (3.5) is replaced by zero. The function which then solves (3.5) has also to satisfy

$$\frac{1}{M} \sum_{z_1 \in S_M^+} \Theta(z)\ F(z) = <\Phi_- | \sigma_1^x | \Phi_+> \tag{3.9}$$

Of particular interest is the $M \to \infty$ limiting behaviour of this set-up. First we establish bound on the $F_M(z)_n$ using the B.C.S. inequality.

$$|F_M(e^{i\alpha})_n|^2 \le \langle\Phi_+|G^+(\alpha_1) \ldots G^+(\alpha_n)\, G(\alpha_n) \ldots G(\alpha_n)|\Phi_+\rangle \tag{3.10}$$

to which Wick's theorem may be applied using the linear dependence (3.3); the r.h.s. of (3.10) is a Pfaffian which is the square root of an antisymmetric determinant [Caianiello, 1953], to which the Hadamard bound may be applied giving

$$\sup_{(z)_n \in K^n} |F_M(z)^n| \le J(K_1,K_2)^n\,(2n)^{n/2} \tag{3.11}$$

for non-critical theories where the compact set K in the domain of analyticity of Θ includes the unit circle.

By Montel's theorem for n-complex variables [Nachbin 1970], there is at least one subsequence $\delta(M) \subset \mathbb{Z}$ on which the $F_M(z)_n$ converge uniformly to $F_\delta((z)_n)$ which is analytic inside K^n. On the intersection of the δ for each side of (3.5) a standard complex variable argument gives

$$(Y_+F)(z)_n = \sum_2^n (-1)^j\, h(z_1,z_j)\, F(\Delta_{ij}(z)_n) \tag{3.12}$$

with

$$(Y_+F)(z)_n = \frac{P}{\pi i}\int \frac{dt}{t}\,\frac{1}{t/z_1-1}\left(1 + \frac{\Theta(t)}{\Theta(z_1)}\right) F(t,\Delta_1(z)_n) \tag{3.13}$$

We shall have a recurrence relation on the limiting matrix elements if we can invert (3.12); this looks like a Fredholm alternative problem [Smithies, 1965; Reed and

Simon, 1971], but since Y_+ is not in general a compact operator, we need a different setting. It is straightforward to show that Y_+ is bounded and self-adjoint on $L^2(S_1)$ (think of $n = 1$ in 3.13). If the null space, denoted B_0, of Y_+ is empty, (3.12) has a unique solution. Suppose, on the other hand, that B_0 is not empty; then (3.12) has no solution unless

$$\sum_{2}^{n} (-1)^j (v,h)(z_j)\, F(\Delta_{1j}(z)_n) = 0 \tag{3.14}$$

for all $v \in B_0$. For either (1,1) or (1,0) transfer we can get the spectrum of Y_+ analytically by a conformal mapping which makes Y_+ a difference kernel with a range of integration suitable for the immediate application of Fourier methods. It is more appealing, though, to use the theory of semi-infinite Toplitz forms [for a review, see Widom, 1965], with result that

$$\dim B_0 = |I(\Theta)| \tag{3.15}$$

In section 2, we saw that this index was also crucial in determining the vacuum structure. For $k > 1$ (high temperature) $\dim B_0 = 0$ from (3.15) and from the vacuum structure, only matrix elements with n even are non-vanishing there is an initial condition

$$F(\phi) = \lim_{M\to\infty} \langle \Phi_- | \Phi_+ \rangle \tag{3.16}$$

(subsequential convergence is assured at least by boundedness). Thus we have

$$F_\infty((e^{i\omega})_{2n}) = \sum_{2}^{2n} (-1)^j\, f_-(\omega_1,\omega_j)\, F_\infty(\Delta_{1j}(e^{i\omega})_{2n}) \tag{3.17}$$

where $f_-(\omega_1,\omega_2)$ is antisymmetric; it may be found by intelligent guesswork or by Wiener-Hopf factorisation [Abraham, 1978a]. The matrix elements are therefore

Pfaffians [Caianiello, 1953]. Strictly speaking, we should retain a notation which express the subsequence for convergence in (3.16). This is somewhat pedantic, and will be dealt with in the next section.

The low-temperature phase has $k < 1$ and thus $\dim B_0 = 1$; the vacuum structure gives non-vanishing matrix elements only for odd n . The null-vector which also satisfies the limit of (3.9) which is

$$\frac{1}{2\pi i} \oint \frac{dz}{z} \quad \Theta(z)\, F(z) \quad = \quad m^* \tag{3.18}$$

may be obtained by Wiener-Hopf methods [Krein, 1958]. This solution when inserted in the (3.14) gives the new criterion

$$\sum_{2}^{2n+1} (-1)^j \; F(z_j)\; F(\Delta_{1j}(z)_{2n+1}) \quad = \quad 0 \tag{3.19}$$

which allows us to invert (3.12) by solving

$$(Y_+ f_-)(z_1, z_2) \quad = \quad h(z_1, z_2) + A\, F(z_1)\; F(z_2) \tag{3.20}$$

with a suitable choice of A ; this can again be done by the Wiener-Hopf technique giving after a little rearrangement

$$\begin{aligned} F((z)_{2n+1}) \quad &= \quad \sum_{1}^{2n+1} (-1)^j \; F(z_j)\; F(\Delta_j(z)_{2n+1}) \\ &+ \quad F(z_1)\; G(\Delta_1(z)_{2n+1}) \end{aligned} \tag{3.21}$$

where G is arbitrary and

$$F(\;)_{2n}) \quad = \quad \sum_{2}^{2n} (-1)^j \; f_-(\omega_1, \omega_j)\; F\;(\Delta_{1j}(\omega)_{2n}) \tag{3.22}$$

with $F(\phi) = 1$. The situation is saved by bringing in the antisymmetry of $F((z)_{2n+1})$ which requires G to vanish.

What remains to be carried out is the evaluation of the initial conditions for the Pfaffian recurrences. In the next section, we shall see how these conditions follow from the expressions for the pair correlation function analysed for coincident points.

Index theorems have been encountered elsewhere [Lewis and Sisson 1975, Lewis and Winninck 1979, Araki and Evans 1983, Evans and Lewis 1986, Carey and Evans 1986], in the description of equilibrium states in a broad C*-algebraic sense rather than the detailed investigation pursued here.

4. Pair correlation function

To get the pair correlation functions in (1.9) the matrix element functions

$$F_M^x((e^{i\alpha})_n) = M^{n/2} \exp i \sum_1^n (\alpha_j+\theta(\alpha_j)) \langle \Phi_- | G(\alpha_n)\ldots G(\alpha_i)\sigma_1^x | \Phi_+ \rangle \tag{4.1}$$

are needed as $M \to \infty$. Such quantities are easily obtained using (2.11) and (2.12) to replace σ_1^x by $f_1^+ + f_1$, followed by (2.26) and (2.27) to relate the result to the matrix elements $F_M((e^{i\alpha})_n)$ investigated in the previous section. The results are [Abraham 1978b]

low-temperature $(k < 1)$: $F_\infty^x((e^{i\omega})_{2n+1}) = 0$ and

$$F_\infty^x((e^{i\omega})_{2n}) = m^* \sum_2^{2n} (-1)^j f_-(\omega_1,\omega_j) F(\Delta_{1j}(\omega)_{2n}) \tag{4.2}$$

with initial condition $F(\phi) = 1$

high temperature $(k > 1)$: $F_\infty^x((e^{i\omega})_{2n}) = 0$ and

$$F_\infty^x((e^{i\omega})_{2n+1}) = \sum_1^{2n+1} (-1)^j g(\omega_j) F(\Delta_j(e^{i\omega})_{2n+1}) \tag{4.3}$$

where the last term is the usual Pfaffian with initial condition as in (3.16). The reader should consult Abraham

[1978b] for details of the function $g(\omega)$ and $f_-(\omega_1,\omega_2)$ in terms of the Wiener-Hopf factorisations of Θ.

The pair correlation function is given by

$$\langle\sigma(x,y)\,\sigma(0,0)\rangle = \lim_{M\to\infty} \sum_0^\infty \frac{1}{n!M^n} \sum_{(\omega)_n\in\Omega_M^n(-)} |F_M^x(e^{i\omega})_n|^2 \exp - \sum_1^n (|x|\gamma(\omega_j)+i_y\omega_j) \tag{4.4}$$

First note that the series converges uniformly in M for any $(x,y)\in\mathbb{Z}^2$ by the Weierstrass M-test; thus the limit can be taken term-by-term using the convergence results of section 3 to give

$$\langle\sigma(x,y)\,\sigma(0,0)\rangle = \sum_0^\infty \frac{1}{n!(2\pi)^n} \int_{-\pi}^{\pi}\!\!\cdots\!\int d(\omega)_n\, |F_\infty(e^{i\omega})_n|^2 \exp - \sum_1^n (|x|\gamma(\omega_j)+i_y\omega_j) \tag{4.5}$$

Notice that for $T < T_c$ the sum is only over even n. Using the identity $(\mathrm{Pf}D)^2 = \det D$ for antisymmetric D, it is clear that (4.5) is actually a Fredholm determinant, since the odd-n terms vanish (if $D = -D^T$ and D is $n \times n$, n odd, then $\det D = 0$) giving

$$\langle\sigma(x,y)\,\sigma(0,0)\rangle = d(1) \tag{4.6}$$

where

$$d(\lambda) = \sum_0^\infty \frac{(-1)^n}{(2\pi)^n n!} \int_{-\pi}^{\pi}\!\!\cdots\!\int K\begin{pmatrix}\omega_1,\ldots\omega_n\\ \omega_1,\ldots\omega_n\end{pmatrix} d(\omega)_n \tag{4.7}$$

with

$$K\begin{pmatrix}\omega_1\ldots\omega_n\\ \omega_1\ldots\omega_n\end{pmatrix} = \det K \tag{4.8}$$

where

$$K_{ij} = i\,h(\omega_i)\ h(\omega_j)\ f_-(\omega_1,\omega_2) \tag{4.9}$$

with

$$h(\omega) = \exp-(x\gamma(\omega) + i_y\omega)/2 \tag{4.10}$$

A useful reference for the classical Fredholm theory is Smithies [1965]. Since we have

$$<\sigma(x,y)\ \sigma(0,0)> \geq (m^*)^2 > 0 \tag{4.11}$$

by the Griffiths inequalities [Griffiths, 1971 and references therein] and the definition of the low-temperature region, (4.6) gives $d(1) > 0$. Thus we can use resolvent formulae to simplify (4.7); this is rather like the linked cluster expansion in statistical physics. Such an approach gets one into inverse scattering, but that is another story. If $x = y = 0$, we can evaluate the l.h.s. of (4.6) trivially, getting $(m^*)^2$ from (4.7) by obtaining the spectrum of $if_-(\omega_1,\omega_2)$ and evaluating the Fredholm determinant.

The work for $T > T_c$ is slightly more complicated because (4.5) is a sum over odd n. But it turns out that

$$|F^x(e^{i\omega})_{2N+1}|^2 = \Sigma \text{ chain } \det D \tag{4.12}$$

where D is antisymmetric, of even dimension. This is a simple graphical result [Kasteleyn, 1967]. Introducing the coupling parameter as before allows us to rearrange the series for $|\lambda|$ small enough to give

$$<\sigma(x,y)\ \sigma(0,0)>(\lambda) = (h,(1+\lambda\ K^2)h)\ d(\lambda) \tag{4.13}$$

followed by analytic continuation to $\lambda = 1$ since $d(1) \neq 0$. The same idea with $x = y = 0$ suffices to evaluate $F(\phi)$ - and, incidentally, shows that only one subsequence actually occurs.

5. Conclusion

In these notes we have shown that the Jordan-Wigner transformation, which plays a central role in the algebraic development of the statistical mechanics of the planar Ising model, is a natural consequence of the star-triangle relation and that the QISM is very useful in this respect since it provides motivation. Then we outlined the matrix element theory of the pair correlation function and its relationship to Fredholm theory.

The author thanks Professor D.E. Evans very much for the hospitality generously extended to him at the operator algebra symposium in Warwick.

References

Abraham, D.B. (1978a). Odd Operators and Spinor algebras in lattice statistics: n-point functions for the rectangular Ising model. Commun. Math. Phys. 59, 17-34.

Abraham, D.B. (1978b). Pair function for rectangular Ising ferromagnet. Commun. Math. Phys. 60, 181-191.

Abraham, D.B. (1978c). n-point functions for the rectangular Ising ferromagnet. Commun. Math. Phys. 60, 205-213.

Abraham, D.B. and Davies, B. (1988). The quantum inverse method and Bogoliubov-Valatin transformation. To be published.

Abraham, D.B. and Martin-Lof, A. (1973). The transfer matrix for a pure phase in the two-dimensional Ising model. Commun. Math. Phys. 32, 245-268.

Abraham, D.B. and Reed, P. (1976). Interface profile of the Ising model in two dimensions. Commun. Math. Phys. 49, 35-46.

Aizenman, M. (1979). Instability of Phase coexistence and translational invariance in two dimensions. Phys. Rev. Letters 43, 407-409.

Aizenman, M. (1980). Translational Invariance and instability of Phase coexistence in the two-dimensional Ising system. Commun. Math. Phys. 73, 83-94.

Araki, H. and Evans D.E. (1983). A C*-algebra approach to phase transitions in the two dimensional Ising model. Commun. Math. Phys. 91, 489-503.

Baxter, R.J. (1982). Exactly solved models in statistical mechanics. London: Academic Press.

Caianiello, E.R. (1953). Combinatorics and renormalisation in quantum field theory. New York: Benjamin.

Carey, A.L. and Evans, D.E. (1986). The Operator Algebras of the two dimensional Ising model. Contemporary Maths. Artin s Braid Group (to appear).

Evans, D.E. and Lewis J.T. (1986). On a C*-algebra approach to phase transitions in the two dimensional Ising model II. Commun. Math. Phys. 102, 521-535.

Gallavotti, G. (1972). The Ising model. Riv. Nuovo Cimento 2, 133.

Griffiths, R.B. (1971). Phase Transitions in Statistical Mechanics and Quantum Field Theory. ed. C. De Witt and R. Stora. pp 241-281. New York: Gordon and Breach.

Gutkin, E. (1986). A comment on Baxter condition for commutativity of transfer matrices. J. Statistical Phys. 44, 193-202.

Higuchi, Y. (1979. On the absence of non-translationally-invariant Gibbs states for the two-dimensional Ising model. Colloq. Math. Soc. Janos Bolyai 27, 517-533.

Kasteleyn, P.W. (1967). Graph theory and crystal physics, in Graph Theory and Theoretical Physics. ed. F. Harary. London: Academic Press.

Kaufman, B. (1949). Crystal Statistics II: partition function evaluated by spinor analysis. Phys. Rev. 76, 1232-1243.

Krein, M.G. (1958). Integral equations on a half-line with kernel depending upon the difference of the arguments. Usp. Mat. Nauk. (N.S.) 13, no.5(83), 3-120. English translation: A.M.S. translations, series 2, 22, 163 (1962).

Lewis, J.T. and Sisson, P.N.M. (1975). A C*-algebra of the two-dimensional Ising model. Commun. Math. Phys. 44, 279-292.

Lewis, J.T. and Winninck, M. (1979). The Ising model phase transition and the index of states of the Clifford algebra. Colloq. Math. Soc. Janos Bolyai 27, 671-679.

Lieb, E.H. and Wu, F.Y. (1972). Two-dimensional Ferroelectric models in Phase Transitions and critical phenomena I, eds. C. Domb and M.S. Green. New York: Academic Press.

Nachbin, L. (1970). Holomorphic Functions, domains of holomorphy and local properties. Math. Studies 1. Amsterdam: North Holland.

Onsager, L. 1944). Crystal Statistics I: a two-dimensional model with an order-disorder transition. Phys. Rev. 65, 117-149.

Onsager, L. (1971). in Critical Phenomena in alloys, magnets and superconductors. eds. R.E. Mills, E. Ascher and R.I. Jaffee. New York: McGraw-Hill.

Onsager, L. and Kaufman, B. (1949). Crystal Statistics III. Short range order on the binary Ising lattice. Phys. Rev. 76, 1244-1252.

Reed, M. and Simon, B. (1971). Methods of Modern Mathematical Physics. I: Functional analysis. New York: Academic Press.

Schultz, T.D., Mattis, D.C. and Lieb, E.H. (1964). Two-dimensional Ising model as a solvable problem of many fermions. Rev. Mod. Phys. 36, 856-871.

Smithies, F. (1965). Integral Equations. Cambridge Tracts in Mathematics and Mathematical Physics 49, Cambridge: Cambridge University Press.

Stephen, M.J. and Mittag, L. (1972). A new representation of the solution of the Ising model. J. Math. Phys. 13, 1944-1951.

Takhtadzhan, L.A. and Fadeev, L.D. (1979). The quantum method of the inverse problem and the Heisenberg XYZ model. Uspekhi Mat. Nauk. 34 : 5, 13-63. English translation: Russian Math. Surveys 34 : 5, 11-68.

Thacker, H.B. (1981). Exact integrability in quantum field theory and statistical systems. Rev. Mod. Phys. 53, 253-285.

Widom, H. (1965). Toplitz matrices. M.A.A. Stud. Math. 3, 179. (ed. I.I. Hirschman).

THE HEAT SEMIGROUP, DERIVATIONS, AND REYNOLDS'S IDENTITY

Charles J.K. Batty[1], Ola Bratteli[2] and Derek W. Robinson[3]
Mathematics Research Centre, University of Warwick

1. Introduction

Let **A** be a C^*-algebra, and **g** a finite dimensional Lie algebra acting as a Lie algebra of closed *-derivations on **A**. We will consider the problem of when **g** exponentiates to a representation of the simply connected covering Lie group G of **g**, as a group of *-automorphisms of **A**, [JM1], [Rob 1], [BGJR1].

Let $\mathbf{A}_\infty$ be the *-algebra of C_∞-elements for **g**, i.e. $\mathbf{A}_\infty$ is the intersection of the domains of all monomials in the elements of **g**. A minimal requirement for **g** to exponentiate is that $\mathbf{A}_\infty$ be dense in **A**, and we will assume this throughout this paper. Recall also that an operator δ is called conservative if δ and $-\delta$ are dissipative, i.e. if $\|(1 + \lambda\delta)(x)\| \geq \|x\|$ for all $x \in D(\delta)$ and all $\lambda \in \mathbf{R}$, [BR1]. Let $\delta_1, ..., \delta_d$ be a basis for **g**.

It was proved in [BGJR1] in a general Banach space setting that **g** exponentiates to an isometric representation of G if and only if the following three conditions hold:

(i) The operators $\delta_1, ..., \delta_d$ are conservative.

(ii) The closure $\overline{\Delta}$ of the Laplacian $\Delta = -\sum_{k=1}^{d} \delta_k^2$ exists as a densely defined operator such that $\overline{\Delta}$ generates a contraction semigroup $S_t = \exp(-t\overline{\Delta})$ on **A** and

$$S_t \mathbf{A} \subseteq \mathbf{A}_\infty$$

for $t > 0$.

(iii) There exists a constant $C > 0$ such that

$$\|\delta_k S_t\| \leq Ct^{-1/2} \tag{1}$$

for $0 < t \leq 1$ and $k = 1, ..., d$.

Note in passing that the conditions of the above Theorem imply that S is a holomorphic semigroup [BGJR1], [Paz 1]. If, on the other hand, it is known a priori that S is a holomorphic semigroup, then condition 1 is equivalent to

---

Permanent addresses: 1. St. John's College, Oxford 0X1 3JP, England.
2. Institute of Mathematics, University of Trondheim, N-7034 Trondheim - NTH, Norway.
3. Department of Mathematics-RSPS, Australian National University-IAS, GPO Box 4, Canberra, ACT 2601, Australia.

$$\|\delta_k x\| \leq C\|x\|^{1/2} [\max \{\|\overline{\Delta}x\|, \|x\|\}]^{1/2} \tag{2}$$

for a possibly different constant C. This will be discussed in section 6.

In some Banach space settings it is known that the "if" part of the above theorem is true under seemingly weaker hypotheses. Most notably, if $\mathbf{A}$ is a Hilbert space and δ_k are skew adjoint operators, then the estimate (1) is automatically true, [Nel 1], [BGJR1]. In this case Δ is already closed, $\overline{\Delta} = \Delta$, and conditions (i), (ii), (iii) reduce to the requirement that Δ is self-adjoint.

The starting point of this paper was originally the desire to weaken the conditions (i), (ii), (iii) in the context of C^*-algebras. We first prove in a general Banach space setting that if $\delta_1, ..., \delta_d$ are conservative, Δ is closed and Δ generates a holomorphic semigroup, then (1) **holds and g exponentiates** to a group of isometries if $e^{-t\Delta}\mathbf{A} \subset \mathbf{A}_\infty$. This is done in Theorem 2.1.

Unfortunately this theorem is uninteresting for C^*-algebras when $d > 1$ in that the hypothesis that Δ is closed is not true in interesting circumstances. The hypotheses of the theorem are equivalent to estimates of the form

$$\|\delta_k^2 x\| \leq C(\|\Delta x\| + \|x\|) ,$$

see Remark 2.2. However, in the case $\mathbf{A} = C_0(\mathbf{R}^d)$, $\mathbf{g} = \mathbf{R}^d$, $\delta_k = \dfrac{\partial}{\partial x_k}$ it can be proved that if Q is any homogeneous differential operator of order 2 in δ_k with constant coefficients, and Q satisfies the estimate above in lieu of δ_k^2, then Q is a constant multiple of Δ, [LM 1], [LM 2].

For abelian C^*-algebras we have better results: If $\mathbf{A}$ is an abelian C^*-algebra and $-\Delta$ is dissipative then $\delta_1, ..., \delta_d$ are conservative. An immediate corollary is that if $d = 1$ and $\mathbf{A}$ is abelian, and $\overline{\Delta}$ is the generator of a contraction semigroup, then $\overline{\delta}_1$ is a generator of a group of $*$-automorphisms, and a posteriori $\overline{\Delta} = -(\overline{\delta}_1)^2$. This is discussed in section 3.

To prove a similar result for non-abelian C^*-algebras $\mathbf{A}$, put $\delta = \delta_1$ and assume that δ is closed and $-\delta^2$ (without closure) generates a contraction semigroup. This entails that the resolvent $(1 - \varepsilon^2 \delta^2)^{-1}$ exists and has norm 1 for $\varepsilon \in \mathbf{R}$. As $1 - \varepsilon^2 \delta^2 = (1 + \varepsilon\delta)(1 - \varepsilon\delta)$ it follows that $1 + \varepsilon\delta$ is $1 - 1$ and the range of $1 + \varepsilon\delta$ is $\mathbf{A}$. As δ is closed, it follows that $(1 + \varepsilon\delta)^{-1}$ exists as a bounded operator for $\varepsilon \in \mathbf{R}$. Also $\|(1 + \varepsilon\delta)^{-1} (1 - \varepsilon\delta)^{-1}\| = \|(1 - \varepsilon^2 \delta^2)^{-1}\| = 1$. If we could conclude that $\|(1 + \varepsilon\delta)^{-1}\| = 1$, δ would be a generator.

We will show in Example 4.4 that the existence of the resolvent $R = (1+\varepsilon\delta)^{-1}$ as a bounded operator does not in itself suffice for the estimate $\|R\| \leq 1$. It has been pointed out in [Rot 1], [Her 1] that the derivation property of δ is equivalent to the Reynolds identity

$$R(xy) = R(x) R(y) + R((1-R)(x)(1-R)(y))$$

for R, and this is a straightforward computation. Since $1 - R = \varepsilon\delta R$ the last term in this identity could be viewed as an $O(\varepsilon^2)$-perturbation of the first; this indicates that $R(xx^*) = R(x) R(x)^* + O(\varepsilon^2)$ and hence that R is positive, and thus has norm one. However, since the derivation in Example 4.4 has empty spectrum, this idea is fallacious. Iterating Reynolds's formula we get

$$R(xy) = \sum_{k=0}^{m-1} (R(1-R)^k x)(R(1-R)^k y) + R\big(((1-R)^m x)((1-R)^m y)\big)$$

for $m = 1, 2, \ldots$. If δ is a generator, we will show that

$$\lim_{m\to\infty} \|(1-R)^m x\| = 0$$

for each $x \in \mathbf{A}$, and hence we get the expansion

$$R(xy) = \sum_{k=0}^{\infty} (R(1-R)^k x)(R(1-R)^k y) ,$$

where the sum converges in norm. Putting $y = x^*$ this gives a proof that R is positive. We will show for a general derivation δ that if $\|(1-R)^k\|$ is uniformly bounded, then this expansion converges and hence R is positive and has norm 1. This is part of Theorem 4.1. However, Reynolds's identity as an algebraic identity alone does not suffice for this: There exists a *-linear operator R on a unital C^*-algebra $\mathbf{A}$ such that $R(1) = 1$, R satisfies Reynolds's identity, and $\|(1-R)^k\|$ is uniformly bounded, but R is not positive (and then $\|R\| > 1$). The example is a projection, i.e. $R^2 = R$, such that R is a conditional expectation in the sense $R(xyz) = xR(y)z$ for $y \in \mathbf{A}$, $x,z \in R(\mathbf{A})$, but R does not have norm 1, see Example 4.3.

To get further insight into Reynolds's identity, we finally show that if R is a positive normal operator satisfying Reynolds's identity on a von Neumann algebra $\mathbf{M}$ with a separating and cyclic vector such that the corresponding state ω is R-invariant, then R decomposes as $R = (1-\delta)^{-1}E$ where E is a normal conditional expectation (of norm 1) from $\mathbf{M}$ onto a von Neumann subalgebra $\mathbf{N}$, and δ is a σ-weakly closed, σ-weakly densely defined *-derivation of $\mathbf{N}$. This is Theorem 5.1, which is a straightforward generalization of Gian-Carlo Rota's corresponding theorem when $\mathbf{M}$ is abelian, [Rot 1].

Example 1.1. (This approach was suggested by David E. Evans, and the counterexample was provided by Man-Duen Choi). Let δ be a derivation on a C^*-algebra $\mathbf{A}$, and assume that $-\delta^2$ (without closure) generate a semigroup of completely positive contractions. Then $(1-\varepsilon\delta^2)^{-1} = \int_0^\infty dt\, e^{-t}\, e^{t\varepsilon^2\delta^2}$ is completely positive. Since $(1+\varepsilon\delta)^{-1}(1-\varepsilon\delta)^{-1} = (1-\varepsilon^2\delta^2)^{-1}$ and $(1-\varepsilon\delta)^{-1} + (1-\varepsilon\delta)^{-1} = 2(1-\varepsilon^2\delta^2)^{-1}$ one could hope that this would imply that $S = (1+\varepsilon\delta)^{-1}$, $T = (1-\varepsilon\delta)^{-1}$ are positive and that δ is a generator by Theorem 4.1. However, there exists a C^*-algebra $\mathbf{A}$ and bounded linear operators S and T on $\mathbf{A}$ such that ST and $S+T$ are completely positive, $ST = TS$, but neither S nor T is positive. An example is $\mathbf{A} = M_2$ with

$$T\left(\begin{bmatrix} a & b \\ c & d \end{bmatrix}\right) = \begin{bmatrix} d & b \\ c & -a \end{bmatrix}$$

and $S = -T$. Then $S + T = 0$, $ST = TS$ and

$$ST\left(\begin{bmatrix} a & b \\ c & d \end{bmatrix}\right) = \begin{bmatrix} a & -b \\ -c & d \end{bmatrix}.$$

Thus ST is an automorphism and hence completely positive. However, S and T are manifestly non-positive.

Note also that if K is any densely defined operator on a Banach space $\mathbf{B}$, then $\delta = \begin{bmatrix} 0 & 0 \\ K & 0 \end{bmatrix}$ is a densely defined operator on $\mathbf{A} = \mathbf{B} \times \mathbf{B}$ such that $\overline{\delta^2} = 0$. Thus it is not true in general that if $-\overline{\delta^2}$ is a generator then δ is so. See [JM1, Example 8.5].

2. Holomorphic semigroups and conservative operators

Theorem 2.1. *Let* $\mathbf{g}$ *be a finite dimensional Lie algebra acting as a Lie algebra of closed conservative operators on a Banach space* $\mathbf{A}$. *Let* $\Delta = -\sum_{k=1}^{d} \delta_k^2$ *be an associated Laplacian. If* Δ *is closed and* Δ *generates a holomorphic semigroup of contractions on* $\mathbf{A}$, *then* $\mathbf{g}$ *exponentiates if and only if* $e^{-t\Delta}\mathbf{A} \subset \mathbf{A}_\infty$ *for* $t > 0$.

Proof. First note that, as in [Rob 1, Lemma 2.3], since δ_k is conservative we have

$$\begin{aligned} \varepsilon\|\delta_k(x)\| &\le \|(1 - \varepsilon\delta_k)(x)\| + \|x\| \\ &\le \|(1 + \varepsilon\delta_k)(1 - \varepsilon\delta_k)(x)\| + \|x\| \\ &= \|(1 - \varepsilon^2\, \delta_k^2)(x)\| + \|x\| \\ &\le \varepsilon^2\|\delta_k^2(x)\| + 2\|x\| \end{aligned}$$

for $\varepsilon > 0$ and $x \in D(\delta_k^2)$. Thus

$$\|\delta_k(x)\| \le \varepsilon\|\delta_k^2(x)\| + \frac{2}{\varepsilon}\|x\|.$$

But the identity $(1 - \varepsilon^2\,\delta_k^2) = (1 + \varepsilon\delta_k)(1 - \varepsilon\delta_k)$ implies that $\|(1 - \varepsilon^2\,\delta_k^2)(x)\| \ge \|x\|$ for $x \in D(\delta_k^2)$, thus δ_k^2 is dissipative and hence closable. As $\Delta = -\sum_{k=1}^{d} \delta_k^2$ is closed and $D(\Delta) \subseteq D(\delta_k^2)$ it follows from Hφrmander's version of the closed graph theorem that there exists a constant $C > 0$ such that

$$\|\delta_k^2(x)\| \le C\big(\|\Delta(x)\| + \|x\|\big)$$

for $x \in D(\Delta)$, [Yos 1, Theorem II.6.2]. Combining this with the previous inequality we get

$$\|\delta_k(x)\| \le \varepsilon C\|\Delta(x)\| + (\varepsilon C + \frac{2}{\varepsilon})\,\|x\|$$

for $x \in D(\Delta)$. But as $e^{-t\Delta} = S_t$ is a holomorphic semigroup, we also have an estimate of the form

$$\|\Delta S_t x\| \le \frac{K}{t}\|x\| \quad \text{for} \quad 0 < t \le 1$$

Thus

$$\begin{aligned}\|\delta_k S_t x\| &\le \frac{\varepsilon}{t} CK\|x\| + (\varepsilon C + \frac{2}{\varepsilon})\|S_t x\| \\ &\le (\frac{\varepsilon}{t} CK + \varepsilon C + \frac{2}{\varepsilon})\|x\| .\end{aligned}$$

Putting $\varepsilon = t^{1/2}$ we get, using $t \le 1$,

$$\|\delta_k S_t x\| \le (CK + C + 2)\frac{1}{t^{1/2}}\|x\|$$

for all $x \in \mathbf{A}$, i.e.

$$\|\delta_k S_t\| \le \frac{CK+C+2}{t^{1/2}}$$

for $k = 1, \dots, d$ and $0 < t \le 1$. But this is the estimate (1), so **g** exponentiates by [BGJR1, Theorem 2.8].

Remark 2.2. Note that if $\delta_1, \dots, \delta_d$ is a finite collection of closed, conservative operators on a Banach space **A** such that $D(\Delta)$ is dense, then the following two conditions are equivalent:

1. *The Laplacian* Δ *is closed.*
2. *There is a constant* $C > 0$ *such that*

$$\|\delta_k^2(x)\| \le C(\|\Delta x\| + \|x\|)$$

for all $x \in D(\Delta)$ and $k = 1, \dots, d$.

The implication 1 => 2 was established in the proof of Theorem 2.1. But the identity $(1 - \varepsilon^2\delta_k^2) = (1 + \varepsilon\delta_k)(1 - \varepsilon\delta_k)$ in conjunction with the estimates $\|(1 - \varepsilon^2\delta_k^2)(x)\| \ge \|x\|$, $x \in D(\delta_k^2)$ and $\|(1 + \varepsilon\delta_k)(y)\| \ge \|y\|$, $y \in D(\delta_k)$, implies that δ_k^2 is actually closed, not only closable. But then it follows from 2. and the definition $\Delta = -\sum_{k=1}^{d} \delta_k^2$ that Δ is closed. See [Rob 1] for a further discussion of 2. and related estimates.

3. Abelian C*-algebras and derivations

Theorem 3.1. *Let* $\mathbf{A} = C_0(\Omega)$ *be an abelian* C^**-algebra with spectrum* Ω , *and* **g** *a finite-dimensional Lie algebra of (not necessarily closed) *-derivations on* **A** *with a basis* $\delta_1, \dots, \delta_d$. *If* $-\Delta = \sum_{k=1}^{d} \delta_k^2$ *is dissipative, then* $\delta_1, \dots, \delta_d$ *are conservative as operators on*

$D(\Delta)$.

Proof. We have to show that if $a \in D(\Delta)$, $a \geq 0$ and $a(\omega) = 0$, then $(\delta_k a)(\omega) = 0$, where ω is an arbitrary point in Ω. To this end, let f be a real polynomial such that $f(0) = 0$ and $f(t) \geq 0$ for $t \geq 0$. Then $f(a) \in D(\delta_k^2)$ and

$$\begin{aligned}\delta_k(f(a)) &= f'(a)\, \delta_k(a),\\ \delta_k^2(f(a)) &= f''(a)\, (\delta_k(a))^2 + f'(a)\, \delta_k^2(a),\end{aligned}$$

and thus

$$\Delta(f(a)) = -f''(a) \sum_{k=1}^{d} (\delta_k(a))^2 + f'(a)\, \Delta(a)\, .$$

Now, as $f(a)(\omega) = f(a(\omega)) = f(0) = 0$ and $f(a) \geq 0$, it follows from dissipativity of $-\Delta$ that

$$\Delta(f(a))(\omega) \leq 0\, ,$$

and the previous relation, evaluated at ω, gives

$$f''(0) \sum_{k=1}^{d} (\delta_k(a)(\omega))^2 - f'(0)\, \Delta(a)(\omega) \geq 0\, .$$

Now, choosing $f(t) = t(1 - nt)^2$ we deduce that

$$-4n \sum_{k=1}^{d} (\delta_k(a)(\omega))^2 - \Delta(a)(\omega) \geq 0$$

for all n, and hence

$$\sum_{k=1}^{d} (\delta_k(a)(\omega))^2 = 0\, .$$

But then $\delta_k(a)(\omega) = 0$ for $k = 1, \ldots, d$.

Corollary 3.2. *If* δ *is a *-derivation on an abelian* C^**-algebra* $\mathbf{A}$ *such that* δ^2 *is closable,* $-\overline{\delta^2}$ *generates a contraction semigroup and* $D(\delta^2)$ *is a core for* δ, *then* δ *is a pregenerator.*

Proof. By Theorem 3.1, $\delta|_{D(\delta^2)}$ is conservative. For $a \in D(\delta^2)$,

$$(1 - \varepsilon^2 \delta^2)(a) = (1 \pm \varepsilon\delta)(1 \mp \varepsilon\delta)(a) \in \overline{(1 \pm \varepsilon\delta)(D(\delta^2))}\, .$$

Since $1 - \varepsilon^2 \delta^2$ has dense range, it follows that $(1 \pm \varepsilon\delta)(D(\delta^2))$ are dense. Thus $\delta|_{D(\delta^2)}$ is a pre-generator, and then δ is a pregenerator since $D(\delta^2)$ is a core for δ.

Example 3.3. There exist an abelian C^*-algebra $\mathbf{A}$, a closed *-derivation δ on $\mathbf{A}$ such that $-\delta^2$ extends to a generator of a holomorphic semigroup of completely positive contractions, but δ has no generator extensions. (δ is necessarily conservative by Theorem 3.1.)

The example is $A = C_0(0, \infty)$ = the C^*-algebra of continuous functions on $(0, \infty)$ vanishing at 0 and ∞, and $\delta = \frac{d}{dx}$ with $D(\delta)$ = the set of absolutely continuous functions $f \in C_0(0, \infty)$ such that $f' \in C_0(0, \infty)$. Then one can verify that δ^2 is closed, with

$$D(\delta^2) = \{f \in C_0(0, \infty) \mid f', f'' \in C_0(0, \infty)\}$$

(To see this, suppose $f_n \in D(\delta^2)$, $\|f_n - f\| \to 0$, $\|f''_n - g\| \to 0$. Take $a > 0$, and let h be a C^2-function on $[0, \infty]$ with compact support such that $h(t) = \int_0^t (t-s)\, g(s)\, ds$ for $0 \le t \le a$. Then $h \in D(\delta^2)$ and $\delta^2(h) = g$ on $[0, a]$. Let $h_n = f_n - h \in D(\delta^2)$. Then $h_n \to f - h$ and $h''_n = \delta^2(h_n) \to g - \delta^2(h)$ uniformly on $[0, a]$. By Taylor's theorem, $h_n(t) = \frac{t^2}{2} h''_n(s_n)$ for $0 < t \le a$ and some $0 < s_n < t \le a$. Letting $n \to \infty$, we deduce that $f(t) - h(t) = 0$ for $0 \le t \le a$. Since a is arbitrary,

$$f(t) = \int_0^t (t-s)\, g(s)\, ds$$

for all $t \ge 0$. Hence

$$f'(t) = \int_0^t g(s)\, ds \to 0$$

as $t \to 0$. Also

$$f''(t) = g(t) \to 0 \quad \text{as} \quad t \to \infty,$$

and $f(t) \to 0$ as $t \to \infty$. Standard manipulations of Taylor's theorem now imply that $f'(t) \to 0$ as $t \to \infty$. Hence $f \in D(\delta^2)$ and $\delta^2(f) = g$.) Now, partial integration shows that

$$\int_0^\infty dt\, e^{-t} (1 - \delta^2)(f)(t) = \int_0^\infty dt\, e^{-t} \big(f(t) - f''(t)\big) = 0$$

for $f \in D(\delta^2)$, hence $(1 - \delta^2)(D(\delta^2))$ is not dense, and $-\delta^2$ is not a generator. However, defining K by $Kf = -f''$ on $D(K) = \{f \in C_0(0, \infty) \mid f'' \in C_0(0, \infty)\}$ then K is an extension of $-\delta^2$, and K is the generator of semigroup given by

$$(e^{-tK} f)(x) = \int_0^\infty p_t(x,y)\, f(y)\, dy$$

for $t > 0$, where

$$p_t(x,y) = \frac{1}{\sqrt{4\pi t}} \left[e^{-\frac{(x-y)^2}{4t}} - e^{-\frac{(x+y)^2}{4t}} \right].$$

This can be seen by extending f to an odd function on $(-\infty, \infty)$ and applying the usual heat semigroup to this extension. On the other hand, δ itself is clearly not a generator, δ is merely the generator of the semigroup of translations to the right. (If $\mathrm{Re}\,\varepsilon > 0$, one calculates that $(1 + \varepsilon\delta)^{-1}$ exists and is given by

$$((1 + \varepsilon\delta)^{-1}f)(x) = \frac{1}{\varepsilon}\int_0^x e^{\frac{t-x}{\varepsilon}} f(t)\, dt ,$$

but if $\mathrm{Re}\,\varepsilon < 0$, one calculates that the range of $(1 + \varepsilon\delta)$ has codimension 1:

$$(1 + \varepsilon\delta)(A) = \{f \in C_0(0, \infty) \mid \int_0^\infty f(t)\, e^{\frac{t}{\varepsilon}}\, dt = 0\} .)$$

Since δ is the generator of a semigroup which can not be extended to a group, δ does not have generator extensions.

4. Derivations

In the following theorem, L_n^m denotes the Laguerre polynomial

$$L_n^m(t) = \frac{e^t\, t^{-m}}{n!}\frac{d^n}{dt^n}(e^{-t}\, t^{n+m})$$

$$= \sum_{j=0}^{n}\binom{n+m}{n-j}\frac{(-t)^j}{j!}$$

for $n = 0, 1, 2, \ldots,$ [EMOT1].

Theorem 4.1. *Let* δ *be a closed *-derivation on a* C^**-algebra* A. *The following eight conditions are equivalent:*

1. δ *is the generator of a strongly continuous semigroup* $t > 0 \mapsto e^{-t\delta}$ *of *-morphisms of* A.

2. $(1 + \varepsilon\delta)^{-1}$ *exists and*

$$\|(1 + \varepsilon\delta)^{-1}\| \leq 1$$

for all $\varepsilon > 0$.

3. $(1 + \varepsilon\delta)^{-1}$ *exists and is positivity preserving for all* $\varepsilon > 0$.

4. $(1 + \varepsilon\delta)^{-1}$ *exists and*

$$\|(\varepsilon\delta(1 + \varepsilon\delta)^{-1})^n\|$$

is uniformly bounded in n *for each* $\varepsilon > 0$.

5. $(1 + \varepsilon\delta)^{-1}$ *exists for all* $\varepsilon > 0$ *and*

$$\|(\varepsilon\delta(1 + \varepsilon\delta)^{-1})^n\| \leq 1 + \int_0^\infty dt\, e^{-t}\, |L_{n-1}^1(t)|$$

for $n = 1, 2, 3, \ldots\ldots$ *Explicitly*

$$\|(\varepsilon\delta(1+\varepsilon\delta)^{-1})^n\| \le \begin{cases} 2 & \textit{for} \quad n = 1 \\ 2(1+e^{-2}) \\ = 2.270670566 \ldots & \textit{for} \quad n = 2 \\ 2[1+(\sqrt{3}-1)e^{-(3-\sqrt{3})} \\ \quad +(\sqrt{3}+1)e^{-(3+\sqrt{3})}] \\ = 2.460140047 \ldots & \textit{for} \quad n = 3 \\ 1 + \sum_{k=0}^{\infty} 1/(k!)^2 \\ = 3.2795 \ldots & \textit{for all} \quad n , \end{cases}$$

where all estimates except the last are the best possible general estimates.

6. $(1 + \varepsilon\delta)^{-1}$ *exists for all* $\varepsilon > 0$ *and*

$$\lim_{n\to\infty} \|(\varepsilon\delta)^n (1 + \varepsilon\delta)^{-n-1}\| = 0 .$$

7. $(1 + \varepsilon\delta)^{-1}$ *exists for all* $\varepsilon > 0$ *and*

$$\lim_{n\to\infty} \|(\varepsilon\delta(1 + \varepsilon\delta)^{-1})^n (x)\| = 0$$

for each $x \in \mathbf{A}$.

8. $R = (1 + \varepsilon\delta)^{-1}$ *exists for all* $\varepsilon > 0$ *and*

$$R(xy) = \sum_{n=0}^{\infty} \left(R(1 - R)^n x\right) \left(R(1 - R)^n y\right)$$

for each pair $x, y \in \mathbf{A}$, *where the latter series converges in norm.*

Furthermore, when the conditions 1-8 are fulfilled, the expansions

$$(\varepsilon\delta(1 + \varepsilon\delta)^{-1})^n(x) = x - \int_0^\infty dt\, e^{-t} L^1_{n-1}(t)\, e^{-t\varepsilon\delta}(x)$$

and

$$(\varepsilon\delta)^n (1 + \varepsilon\delta)^{-n-1}(x) = \int_0^\infty dt\, e^{-t} L^0_n(t)\, e^{-t\varepsilon\delta}(x)$$

are valid, and

$$\|(\varepsilon\delta)^n (1 + \varepsilon\delta)^{-n-1}\| \le \int_0^\infty dt\, e^{-t} \,|L^0_n(t)|\, ,$$

this being the best possible general estimate.

Remark 4.2. During the proof of Theorem 4.1, we will also establish that each of the eight conditions of the theorem are equivalent to each of the following conditions:

4(p). *For each* $\varepsilon > 0$, $(1 + \varepsilon\delta)^{-1}$ *exists and*

$$\|(\varepsilon\delta)^n (1 + \varepsilon\delta)^{-n-p}\|$$

is uniformly bounded in n. ($p = 0, 1, 2, \ldots$.)

6(p). *For each* $\varepsilon > 0$, $(1 + \varepsilon\delta)^{-1}$ *exists and*

$$\lim_{n\to\infty} \|(\varepsilon\delta)^n (1 + \varepsilon\delta)^{-n-p}\| = 0 .$$

($p = 1, 2, 3, \ldots$.)

7(p). *For each* $\varepsilon > 0$, $(1 + \varepsilon\delta)^{-1}$ *exists and*

$$\lim_{n\to\infty} \|(\varepsilon\delta)^n (1 + \varepsilon\delta)^{-n-p}(x)\| = 0$$

for each $x \in \mathbf{A}$. ($p = 0, 1, 2, \ldots$.)

Note in particular that condition 6(0) is not among the equivalent conditions; in fact if δ is an unbounded generator then the spectrum of $\varepsilon\delta(1 + \varepsilon\delta)^{-1}$ contains 1, so $\|(\varepsilon\delta(1 + \varepsilon\delta)^{-1}))^n\| \geq 1$ for all n.

Proof. $1 \Rightarrow 2$ and $1 \Rightarrow 3$ are standard consequences of the Laplace transform formula

$$(1 + \varepsilon\delta)^{-1}(x) = \int_0^\infty dt\, e^{-t}\, e^{-t\varepsilon\delta}(x),$$

see [BR1].

2 => 1: It is immediate from Hille-Yosida theory that δ generates a one-parameter semigroup $e^{-t\delta}$ of contractions on $\mathbf{A}$. To show that these are *-morphisms, one may for example apply the resolvent n-1 times to Reynolds identity to obtain

$$(1 + \frac{t}{n}\delta)^{-n} (xy)$$

$$= (1 + \frac{t}{n}\delta)^{-n}(x)\, (1 + \frac{t}{n}\delta)^{-n}(y)$$

$$+ \frac{t^2}{n^2} \sum_{k=1}^{n} (1+\frac{t}{n}\delta)^{-k} \left((1+\frac{t}{n}\delta)^{-n-1+k}(\delta(x))(1+\frac{t}{n}\delta)^{-n-1+k}(\delta(y))\right)$$

for $x, y \in D(\delta)$, and hence we get the estimate

$$\|(1 + \frac{t}{n}\delta)^{-n}(xy) - (1 + \frac{t}{n}\delta)^{-n}(x)\, (1 + \frac{t}{n}\delta)^{-n}(y)\|$$

$$\leq \frac{t^2}{n} \|\delta(x)\|\, \|\delta(y)\| .$$

Letting $n \to \infty$, we deduce

$$e^{-t\delta}(xy) = e^{-t\delta}(x)\, e^{-t\delta}(y),$$

see [BR1, Theorem 3.1.10]. Thus $e^{-t\delta}$ is a semigroup of *-morphisms.

3 => 2: If $R = (1+\varepsilon\delta)^{-1}$ is positive, it follows from Reynolds's identity

$$R(x^*x) = R(x)^* R(x) + R\Big(((1-R)x)^* \, ((1-R)x)\Big)$$

that the generalized Schwarz's inequality

$$R(x)^* R(x) \le R(x^*x)$$

is valid, thus $\|R(x)\| \le \|R\|^{1/2} \|x\|$ and $\|R\| \le \|R\|^{1/2}$, hence $\|R\| \le 1$. (Incidentally, this means that condition (A1, B1, C2) can be included in Theorem 3.2.50 of [BR1], even when **A** does not have an identity).

This shows the equivalence of 1, 2 and 3.

1 => 5: Replacing δ by $\varepsilon\delta$ we may and will assume that $\varepsilon = 1$. By applying the Laplace transform n times we have

$$(1+\delta)^{-n}(x) = \int_0^\infty dt\, e^{-t} \frac{t^{n-1}}{(n-1)!} e^{-t\delta}(x)$$

Thus assuming for the moment that $x \in D(\delta^n)$ we have, using partial integration:

$$\begin{aligned}
\delta^n(1+\delta)^{-n}(x) &= \int_0^\infty dt\, e^{-t} \frac{t^{n-1}}{(n-1)!} \delta^n e^{-t\delta}(x) \\
&= \int_0^\infty dt\, e^{-t} \frac{t^{n-1}}{(n-1)!} (-\frac{d}{dt})^n e^{-t\delta}(x) \\
&= \int_0^\infty dt\, (\frac{d}{dt})^{n-1} \left[e^{-t} \frac{t^{n-1}}{(n-1)!}\right] (-\frac{d}{dt})\, d^{-t\delta}(x) \\
&= x + \int_0^\infty dt\, (\frac{d}{dt})^n \left[e^{-t} \frac{t^{n-1}}{(n-1)!}\right] e^{-t\delta}(x) \\
&= x + \int_0^\infty \frac{n}{t} e^{-t} L_n^{-1}(t)\, e^{-t\delta}(x)\,.
\end{aligned}$$

But as $L_n^{-1}(t) = -\frac{t}{n} L_{n-1}^1(t)$, [EMOT1], we get

$$\delta^n(1+\delta)^{-n}(x) = x - \int_0^\infty dt\, e^{-t} L_{n-1}^1(t)\, e^{-t\delta}(x)$$

for $x \in D(\delta^n)$. This relation for general $x \in$ **A** follows by closure. (Alternatively, one could deduce this relation by using

$$\delta^n(1+\delta)^{-n} = \left(1 - (1+\delta)^{-1}\right)^n = \sum_{k=0}^n (-1)^k \binom{n}{k} (1+\delta)^{-k}\,.)$$

The estimate

$$\|(\delta(1+\delta)^{-1})^n\| \leq 1 + \int_0^\infty dt\, e^{-t}\, |L_{n-1}^1(t)|$$

is now immediate, and in the case that $\mathbf{A} = C_0(\mathbb{R})$, $\delta = \frac{d}{dx}$, the norm of $(\delta(1+\delta)^{-1})^n$ is actually equal to the right hand side.

The estimates on $\|(\delta(1+\delta)^{-1}))^n\|$ for $n = 1, 2, 3$ are now straightforward. For the general estimate, we again use $L_{n+1}^{-1}(t) = -\frac{t}{n+1} L_n^1(t)$ to obtain

$$\begin{aligned}
&\int_0^\infty dt\, e^{-t}\, |L_n^1(t)| \\
&= \int_0^{\frac{1}{n+1}} dt\, e^{-t}\, |L_n^1(t)| + \int_{\frac{1}{n+1}}^\infty dt\, \frac{n+1}{t}\, e^{-t}\, |L_{n+1}^{-1}(t)| \\
&< \int_0^{\frac{1}{n+1}} dt\, |L_n^1(t)| + \frac{n+1}{(n+1)^{1/2}} \int_{\frac{1}{n+1}}^\infty dt\, e^{-t}\, \frac{1}{t^{1/2}}\, |L_{n+1}^{-1}(t)| \\
&< \int_0^{\frac{1}{n+1}} dt\, |L_n^1(t)| + (n+1)^{1/2} \Big(\int_0^\infty dt\, e^{-t}\, t^{-1}\, |L_{n+1}^{-1}(t)|^2\Big)^{1/2} \\
&= I_1 + I_2\, ,
\end{aligned}$$

where we used Schwarz's inequality.

We now estimate I_1 and I_2 separately: Using

$$L_n^1(t) = \sum_{k=0}^{n} \binom{n+1}{n-k} \frac{(-t)^k}{k!}\, ,$$

we have

$$\begin{aligned}
I_1 = \int_0^{\frac{1}{n+1}} dt\, |L_n(t)| &\leq \sum_{k=0}^{n} \binom{n+1}{n-k} \frac{1}{(k+1)!} \frac{1}{(n+1)^{k+1}} \\
&\leq \sum_{k=0}^{n} \frac{1}{((k+1)!)^2}\, .
\end{aligned}$$

By the orthogonality relations for Laguerre polynomials,

$$\int_0^\infty dt\, e^{-t}\, t^{-1}\, |L_{n+1}^{-1}(t)|^2 \;=\; \frac{n!}{(n+1)!} \;=\; \frac{1}{n+1}\,,$$

and hence

$$I_2 \;=\; 1\,.$$

Thus

$$\int_0^\infty dt\, e^{-t}\, |L_n^1(t)| \;<\; \sum_{k=0}^\infty \frac{1}{(k!)^2}\,,$$

and this gives the last inequality in 5.

5 => 4 is trivial.

4 => 4(p) for $p = 0, 1, 2, \ldots$ is trivial.

4(p) => 6(p+1) for $p = 0, 1, 2, \ldots$:

Again we may assume $\varepsilon = 1$. Put $R = (1+\delta)^{-1}$ and $T = \delta(1+\delta)^{-1} = 1 - R$. Then condition 4(p) implies that $\|T^n R^p\|$ is uniformly bounded in n . Thus the series

$$(1 - \frac{T}{\lambda})^{-1} R^p \;=\; \sum_{n=0}^\infty \frac{T^n}{\lambda^n} R^p$$

converges provided $|\lambda| > 1$, where the left side of the above expression so far has only formal significance. But $R = 1 - T$, so formally

$$\begin{aligned}(1 - \frac{T}{\lambda})^{-1} R^p \;&=\; (1 - \frac{T}{\lambda})^{-1} (1 - \lambda + \lambda(1 - \frac{T}{\lambda}))^p \\ &=\; (1-\lambda)^p (1 - \frac{T}{\lambda})^{-1} + \sum_{k=1}^p \binom{p}{k} (1-\lambda)^{p-k} \lambda^k (1 - \frac{T}{\lambda})^{k-1}\end{aligned}$$

which gives

$$(1 - \frac{T}{\lambda})^{-1} \;=\; \frac{1}{(1-\lambda)^p} \sum_{n=0}^\infty \frac{T^n}{\lambda^n} R^p - \sum_{k=1}^p \binom{p}{k} \left(\frac{\lambda}{1-\lambda}\right)^k (1 - \frac{T}{\lambda})^{k-1}$$

It is clear that algebraic manipulations will show that the expression to the right is the resolvent $(1 - \frac{T}{\lambda})^{-1}$ for $|\lambda| > 1$, hence the spectrum of T is contained in the unit circle.

Repeating this argument, with δ replaced by $\varepsilon\delta$ for $\varepsilon > 0$, it follows that if z is contained in the spectrum of δ , then

$$\left| \frac{\varepsilon z}{1 + \varepsilon z} \right| \;\le\; 1$$

for all $\varepsilon > 0$, i.e. $\operatorname{Re} z \ge 0$. But the set of $\frac{z}{1+z}$ such that $\operatorname{Re} z \ge 0$ is the disc of radius ½ around ½ , thus the spectrum of T is contained in this disc. In particular the spectrum of T is contained in the unit disc, and intersects the unit circle only in 1. Now

$$(1-T)\,(1-zT)^{-1}\,R^p = \sum_{n=0}^{\infty} (T^n - T^{n+1})\,R^p\,z^n\,,$$

the power series having radius of convergence 1 and singular set $\{1\}$. Furthermore

$$\|\sum_{n=0}^{N} (T^n - T^{n+1})\,R^p\| = \|(1-T^{n+1})\,R^p\|\,,$$

which is uniformly bounded in n. It follows from [AOR, Theorem 4] that $\lim_{n\to\infty} \|(T^n - T^{n+1})\,R^p\| = 0$, which is 6(p+1). Note that if $p = 0$, this implication also follows from Katznelson-Tzafriri's stability theorem, [KT1, Theorem 1 with Remark], which states that if T is a linear operator on a Banach space such that $\|T^n\|$ is uniformly bounded in n, then $\lim_{n\to\infty} \|T^n - T^{n+1}\| = 0$ if (and only if) the spectrum of T intersects the unit circle in at most the point $z = 1$.

6(p) => 7(p) for $p = 1, 2, 3 \ldots$ is trivial.

4 => 7 (= 7(0)): As in the proof of 4 => 6 (i.e. 4(0) => 6(1)) above one, uses Katznelson-Tzafriri's stability theorem [KT1], [AOR1] to argue that $\lim_{n\to\infty} \|T^n(1-T)\| = 0$, but as Range $(1-T) = D(\delta)$, it follows that

$$\lim_{n\to\infty} \|T^n x\| = 0$$

for each $x \in D(\delta)$. Since $D(\delta)$ is dense, it follows from 4 that this holds for each $x \in \mathbf{A}$, and that is 7. (In this argument we could also have used Arendt-Batty's stability theorem, [AB1, Theorem 2.4 and Remark 3.3], but the latter is formulated in the setting of one-parameter semigroups, and is more general than we need).

7 => 8: Since $\varepsilon\delta(1+\varepsilon\delta)^{-1} = 1 - R$, it is clear from 7 that the remainder term in the iterated Reynolds's formula

$$R(xy) = \sum_{k=0}^{n-1} (R(1-R)^k x)\,(R(1-R)^k y)$$

$$+ R\big(((1-R)^n x)\,((1-R)^n y)\big)$$

tends to zero as $n \to \infty$. This proves 8.

8 => 3: It is clear from the expansion 8 that $R(xx^*) \geq 0$ for all $x \in \mathbf{A}$, so R is positivity preserving.

7(p) => 3: It follows from 7(p) that $(1-R)^n(x)$ tends to zero for each $x \in$ Range (R^p). As in 7 => 8 it follows that

$$R(xx^*) = \sum_{k=0}^{\infty} (R(1-R)^k x)\,(R(1-R)^k x)^* \geq 0$$

for $x \in \text{Range}(R^p)$. Since $\text{Range}(R) = D(\delta)$, which is dense, $\text{Range}(R^p)$ is also dense, so R is positivity preserving by closure.

This ends the proof of the equivalences in Theorem 4.1. The expansion of $\varepsilon\delta\left(1+\varepsilon\delta)^{-1}\right)^n$ was obtained in the proof of 1 => 5, and similarly

$$\begin{aligned}\delta^n(1+\delta)^{-n-1}(x) &= \int_0^\infty dt\, e^{-t}\, \frac{t^n}{n!}\, \left(-\frac{d}{dt}\right)^n e^{-t\delta}(x)\\ &= \int_0^\infty dt \left(\frac{d}{dt}\right)^n \left(e^{-t}\, \frac{t^n}{n!}\right) e^{-t\delta}\, dt\\ &= \int_0^\infty dt\, e^{-t}\, L_n^0(t)\, e^{-t\delta}\,(x)\,.\end{aligned}$$

Thus the best possible estimate is

$$\|\delta^n(1+\delta)^{-n-1}\| \le \int_0^\infty dt\, |L_n^0(t)|\,,$$

and this estimate is attained for $A = C_0(\mathbf{R})$, $\delta = \dfrac{d}{dx}$.

This ends the proof of Theorem 4.1, but we finally remark that condition 7, $\lim_{n\to\infty} \|\left(\varepsilon\delta(1+\varepsilon\delta)^{-1}\right)^n(x)\| = 0$, can be proved directly, without using Katznelson-Tzafriri's theorem, when δ is the generator of an automorphism group, as follows: If $A[-N, N]$ is the Arveson spectral subspace corresponding to a finite interval $[-N, N]$, then the spectrum of the restriction of $\varepsilon\delta(1+\varepsilon\delta)^{-1}$ to $A[-N, N]$ is a sector of the circle of radius ½ around ½ which does not contain 1. It follows that

$$\lim_{n\to\infty} \|\left(\varepsilon\delta(1+\varepsilon\delta)^{-1}\right)^n |_{A[-N,\, N]}\| = 0\,.$$

But approximating a general element x with an element in some $A[-N, N]$ in norm, and using that $\|\left(\varepsilon\delta(1+\varepsilon\delta)^{-1}\right)^n\|$ is uniformly bounded, it follows that

$$\lim_{n\to\infty} \|\left(\varepsilon\delta(1+\varepsilon\delta)^{-1}\right)^n(x)\| = 0\,.$$

We next show that Reynolds's identity alone, as an algebraic identity, does not suffice for the implication 4 => 3 in Theorem 4.1.

Example 4.3. There exists a C^*-algebra A and a *-linear bounded operator E on A such that E satisfies Reynolds's identity

$$E(xy) = E(x)\,E(y) + E\big((1-E)(x)\,(1-E)(y)\big)\,,$$

and, moreover, $\|(1-E)^n\|$ is uniformly bounded in n, but E is not positivity preserving.

Indeed, if E is any *-linear operator which is a conditional expectation in the sense

$$E^2 = E$$

and

$$E(xy) = xE(y), \quad E(yx) = E(y)x$$

for $x \in E(\mathbf{A})$, $y \in \mathbf{A}$, then E satisfies Reynolds's identity. But a conditional expectation in this sense does not need to be positive. (This was pointed out to us by Jun Tomiyama, and he also provided the following example). Take any C^*-algebra $\mathbf{A}$ with a sub-C^*-algebra $\mathbf{B}$ such that there are more than one projection of norm one from $\mathbf{A}$ onto $\mathbf{B}$. For example, if $\mathbf{A} = \mathbf{L(H)}$ and $\mathbf{B}$ is a maximal abelian subalgebra of $\mathbf{L(H)}$ without minimal projections, then there exists both a normal and a non-normal projection of norm one from $\mathbf{A}$ onto $\mathbf{B}$, [Tom 1]. Let E_1, E_2 be two such projections, and put

$$E = E_1 + \lambda(E_1 - E_2) ,$$

where λ is a real number. As both E_1 and E_2 have the same range one has $E_1 E_2 = E_2$ and $E_2 E_1 = E_1$, and hence E is a conditional expectation in the above sense. But if λ is chosen large (positive or negative) it is clear that E can be made not positivity-preserving. Moreover, as $(1-E)^n = (1-E)$, the sequence $\|(1-E)^n\|$ is uniformly bounded.

We next show that even the resolvent of a derivation need not be positivity preserving:

Example 4.4. There exists a C^*-algebra $\mathbf{A}$ and a densely defined *-derivation δ on $\mathbf{A}$ such that the spectrum of δ is empty, i.e. $(\lambda - \delta)^{-1}$ exists as a bounded operator for all $\lambda \in \mathbb{C}$, but δ is not a generator.

Let $\mathbf{A}$ be the C^*-algebra of continuous functions on $[0, 1]$ vanishing at 0, and let $\delta = \dfrac{d}{dx}$ defined on the set $D(\delta)$ of $f \in \mathbf{A}$ such that $f' \in \mathbf{A}$. The differential equation

$$\lambda f - f' = g$$

where $g(0) = 0$, with the initial condition $f(0) = 0$, has the unique solution

$$f(x) = (\lambda - \delta)^{-1}(g)(x) = -\int_0^x dy\, e^{\lambda(x-y)}\, g(y)$$

for each $\lambda \in \mathbb{C}$. Thus $(\lambda - \delta)^{-1}$ exists, and we have the identity:

$$\|(\lambda - \delta)^{-1}\| = \sup_{0\le x\le 1} e^{\mathrm{Re}(\lambda)x}\int_0^x dy\, e^{-\mathrm{Re}(\lambda)y}$$

$$= \begin{cases} \dfrac{e^{\mathrm{Re}\,\lambda} - 1}{\mathrm{Re}\,\lambda} & \text{if} \quad \mathrm{Re}\,\lambda \neq 0 \\ 1 & \text{if} \quad \mathrm{Re}\,\lambda = 0 \end{cases}$$

We thus see that

$$\|(\lambda - \delta)^{-1}\| \le \frac{1}{|\mathrm{Re}\,\lambda|}$$

if $\mathrm{Re}\,\lambda < 0$, and hence $-\delta$ generates the semigroup $t \ge 0 \Rightarrow e^{-t\delta}$ of translations to the right given by

$$(e^{-t\delta} f)(x) = \begin{cases} f(x-t) & \text{if } \quad x \ge t \\ 0 & \text{otherwise} . \end{cases}$$

Note in particular that $e^{-t\delta} = 0$ for $t \ge 1$. On the other hand

$$\|(\lambda - \delta)^{-1}\| > \frac{1}{\mathrm{Re}\,\lambda}$$

if $\mathrm{Re}\,\lambda > 0$, so δ is not a semigroup generator. This confirms that translation to the left is not well-defined in this example.

Applying the resolvent formula two times, we get

$$\begin{aligned}(\lambda^2 - \delta^2)^{-1}(g)(x) &= (\lambda - \delta)^{-1}\,(\lambda + \delta)^{-1}(g) \\ &= \frac{1}{\lambda}\int_0^x dy\ \sinh\,(\lambda(x-y))g(y)\end{aligned}$$

for $\lambda > 0$. This leads to the identity

$$\|(\lambda^2 - \delta^2)^{-1}\| = \frac{\cosh(\lambda) - 1}{\lambda^2}$$

for $\lambda > 0$, and hence $-\delta^2$ is not a generator of a semigroup in this example, corroborating Corollary 3.2.

Finally, the identity

$$\left(\lambda - (\mu-\delta)^{-1}\right)^{-1} = \frac{1}{\lambda}(1 + \frac{1}{\lambda}(\mu - \frac{1}{\lambda} - \delta)^{-1}) ,$$

valid for $\lambda \ne 0$, shows that the spectrum of $(\mu - \delta)^{-1}$ is $\{0\}$ for all $\mu \in \mathbf{C}$. In particular this means that the spectrum of $-\varepsilon\delta(1 + \varepsilon\delta)^{-1} = 1 - (1 + \varepsilon\delta)^{-1}$ is $\{1\}$ for $\varepsilon > 0$, and the spectral radius formula implies that for any $\eta > 0$ there is a $C_{\eta,\varepsilon} > 0$ such that

$$\|\left(\varepsilon\delta(1-\varepsilon\delta)^{-1}\right)^n\| \le C_{\eta,\varepsilon}(1 + \eta)^n .$$

This shows that the estimate

$$\|\left(\varepsilon\delta(1+\varepsilon\delta)^{-1}\right)^n\| \le C_\varepsilon$$

in condition 4 of Theorem 4.1 cannot be weakened much in the implication 4 => 1.

5. Reynolds's identity

Theorem 5.1. *Let* **M** *be a von Neumann algebra on a Hilbert space* **H** , ω *a vector state* *on* $\mathbf{M}$ *such that the associated vector* Ω *is cyclic and separating, and let* R *be a *-linear bounded operator on* $\mathbf{M}$ *satisfying*

$$R(xy) = R(x)\,R(y) + R\big(((1-R)x)\,((1-R)y)\big)\,,$$

$$R(1) = 1\,,$$

$$\omega\big(R(x)^*\,R(x)\big) \leq \omega(x^*x)\,,$$

for all $x, y \in \mathbf{M}$. *Let* $\mathbf{N}$ *be the* σ*-weak closure of* $R(\mathbf{M})$. *Then there exists a (unique) projection* E *of norm 1 from* $\mathbf{M}$ *onto* $\mathbf{N}$, *and a (unique)* σ*-weakly closed,* σ*-weakly densely defined derivation* δ *on* $\mathbf{N}$ *such that* $(1-\delta)^{-1}$ *exists as a* σ*-weakly continuous operator on* $\mathbf{N}$ *and*

1. $\omega \circ E = \omega,\ \omega \circ \delta = 0\,,$
2. $ER = RE\,,$
3. $R(x) = (1-\delta)^{-1}\big(E(x)\big)$ *for each* $x \in \mathbf{M}$.

Remark 5.2. It is unclear whether the hypotheses of the theorem imply that there exists a semigroup of *-morphisms α of $\mathbf{N}$ such that δ is the generator of α and thus

$$(1-\delta)^{-1} = \int_0^\infty dt\, e^{-t}\,\alpha_t\,.$$

In [Rot 1, Theorem 2] this is claimed to be true when $\mathbf{M}$ is abelian, but the proof has a gap: If H is the skew-symmetric operator implementing δ on $\mathbf{N}\Omega$,

$$Hx\Omega = \delta(x)\Omega$$

for $x \in D(\delta)$, it is clear that $(1-H)^{-1}$ exists and thus H generates a semigroup of contractions $t \geq 0 \to e^{tH}$ on $\overline{\mathbf{N}\Omega}$. It is however not clear a priori that this semigroup leaves the subspace $\mathbf{N}\Omega$ invariant, and this is used in point (8) of the proof in [Rot 1].

Note that the generator of the semigroup

$$x \mapsto (e^{tH})^*x\,e^{tH}$$

of contractions on $\mathbf{L}(\overline{\mathbf{N}\Omega})$ is an extension of $-\delta$, and hence $-\delta$ is dissipative:

$$\|(1+\varepsilon\delta)(x)\| \geq \|x\|$$

for $\varepsilon \geq 0$, $x \in D(\delta)$.

Proof. The inequality $\omega\big(R(x)^*\,R(x)\big) \leq \omega(x^*x)$ immediately implies that there exists an operator T on $\mathbf{H}$ defined by

$$Tx\Omega = R(x)\Omega$$

for $x \in \mathbf{M}$, and

$$\|T\| \leq 1\,.$$

Since $T\Omega = \Omega$ and $\|T\| = 1$, it follows from a theorem of Riesz and Nagy that

$$T^*\Omega = \Omega ,$$

see [RN1]. (Actually, one uses

$$\begin{aligned}\|\Omega\|^2 = (T\Omega, \Omega) = (\Omega, T^*\Omega) &\leq \|\Omega\| \, \|T^*\Omega\| \\ &\leq \|T^*\| \, \|\Omega\|^2 = \|\Omega\|^2\end{aligned}$$

to deduce $\|T^*\Omega\| = \|\Omega\|$, which implies

$$\|\Omega - T^*\Omega\|^2 = \|\Omega\|^2 - (\Omega, T^*\Omega) - (T^*\Omega, \Omega) + \|T^*\Omega\|^2 = 0)$$

In particular, this means

$$\begin{aligned}\omega(R(x)) &= (\Omega, Tx\Omega) = (T^*\Omega, x\Omega) \\ &= (\Omega, x\Omega) = \omega(x) ,\end{aligned}$$

i.e. ω is R-invariant.

We next argue that the operator $2T - 1$ is isometric, i.e.

$$(2T - 1)^* (2T - 1) = 1 .$$

To this end we use Reynolds's identity in the equivalent form

$$R(xR(y) + R(x)y) = R(x) R(y) + R(R(x) R(y)) .$$

for $x, y \in \mathbf{M}$. Applying ω and using $\omega \circ R = \omega$, we get

$$\begin{aligned}&(\Omega, (xR(y) + R(x)y)\Omega) \\ &= (\Omega, R(x) R(y)\Omega) + (\Omega, R(x) R(y)\Omega) \\ &= 2(\Omega, R(x) R(y)\Omega),\end{aligned}$$

and hence

$$\begin{aligned}&(x^*\Omega, Ty\Omega) + (Tx^*\Omega, y\Omega) \\ &= 2(Tx^*\Omega, Ty\Omega),\end{aligned}$$

i.e.

$$((1-2T)x^*\Omega, (1-2T)y\Omega) = (x^*\Omega, y\Omega) ,$$

so $1-2T$ is isometric.

Put $U = 1-2T$. Since U is isometric, Riesz-Nagy's ergodic theorem applies to give an orthogonal projection F on $\mathbf{H}$ such that $1-F$ is the strong limit of the averages $(1 + U + U^2 + ... + U^n) / (n + 1)$. Then

(a) $(1 - F)(\mathbf{H})$ consists of the $\psi \in \mathbf{H}$ such that $U\psi = \psi$.

(b) $F(\mathbf{H}) = \overline{(1-U)(\mathbf{H})} = \overline{T\mathbf{H}}$, and, since $T\Omega = \Omega$, we get $F\Omega = \Omega$.

(c) F commutes with U and thus with T .

Reynolds's identity implies that $R(\mathbf{M})$ is a sub-algebra of $\mathbf{M}$, hence $\mathbf{N} = R(\mathbf{M})''$ is a von Neumann subalgebra of $\mathbf{M}$. Then

$$\overline{\mathbf{N}\Omega} = \overline{R(\mathbf{M})\Omega} = \overline{T\mathbf{M}\Omega} = F(H) .$$

We now argue that there exists a projection of norm one $E : \mathbf{M} \to \mathbf{N}$ with $\omega \circ E = \omega$. By [Tak 1] there exists a (unique) such projection if and only if $\mathbf{N}$ is globally invariant under the modular automorphism group associated to $(\mathbf{M}, \Omega)$, i.e. if and only if the modular operator Δ commutes strongly with F . Let J be the modular conjugation associated to $(\mathbf{M}, \Omega)$ and let $x \in \mathbf{M}$. Then

$$\begin{aligned} J\Delta^{1/2}\, Tx\Omega &= J\Delta^{1/2}\, R(x)\Omega \\ &= R(x)^*\Omega \\ &= R(x^*)\Omega \\ &= Tx^*\Omega \\ &= TJ\Delta^{1/2}x\Omega \end{aligned}$$

from which we get

$$J\Delta^{1/2}T = TJ\Delta^{1/2} ,$$

and hence

$$J\Delta^{1/2}U = UJ\Delta^{1/2} .$$

But as U is an isometry, it follows that U commutes strongly with the components J and $\Delta^{1/2}$ of the polar decomposition of $J\Delta^{1/2}$. Hence T commutes with Δ^{it} for $T \in \mathbb{R}$, and as $F(H) = \overline{T\mathbf{M}\Omega}$ it follows that F commutes with Δ^{it} for $t \in \mathbb{R}$. By Takesaki's theorem the conditional expectation E exists and is unique, and

$$E(x)\Omega = Fx\Omega$$

for $x \in \mathbf{M}$.

We next argue that $R|_{\mathbf{N}}$ has the form $(1 - \delta)^{-1}$ where δ is a σ-weakly densely defined, σ-weakly closed $*$-derivation of $\mathbf{N}$. Since Reynolds's identity is equivalent to the derivation property, it suffices to show that $R(\mathbf{N})$ is dense in $\mathbf{N}$ and that $R|_{\mathbf{N}}$ is $1 - 1$. We first show the latter property. If $x \in \mathbf{N}$ and $R(x) = 0$, then

$$Tx\Omega = R(x)\Omega = 0 .$$

As $x \in \mathbf{N}$ and $F(H) = \overline{\mathbf{N}\Omega}$ it follows that $Fx\Omega = x\Omega$. Hence if $x\Omega \neq 0$, then $(1 - U)x\Omega \neq 0$ by (a), and as $U = 1 - 2T$ we get $Tx\Omega \neq 0$ which contradicts $Tx\Omega = 0$. Thus $x\Omega = 0$ and $x = 0$ since Ω is separating. Thus $R|_{\mathbf{N}}$ is $1 - 1$.

We now argue that $R(\mathbf{N})$ is dense in $\mathbf{N}$, and first show

$$\overline{R(\mathbf{N})\Omega} = \overline{\mathbf{N}\Omega} = F(H) .$$

So let $\psi \in F(H)$ be such that

$$(\psi, R(x)\Omega) = 0$$

for all $x \in N$, i.e.

$$(T^*\psi, x\Omega) = 0$$

for all $x \in N$, i.e.

$$FT^*\psi = 0 .$$

But as $\psi \in F(H)$ and T, and thus T^*, commute with F, it follows that $T^*\psi \in F(H)$ and hence $T^*\psi = 0$. Thus $U^*\psi = (1 - 2T^*)\psi = \psi$. But $\|U^*\| = 1$ so applying Riesz-Nagy's theorem again we have $U\psi = \psi$. But then $\psi \in (1-F)(H)$ and as $\psi \in F(H)$ it follows that $\psi = 0$. Hence, $R(N)\Omega$ is dense in $F(H)$. But as R commutes with the modular automorphism group (since T commutes with Δ^{it}) it follows that $R(N)$ is invariant under this group. As $R(N)\Omega$ is dense in $N\Omega$, it follows that $R(N)$ is dense in N.

As remarked above, there exists a σ-densely defined σ-weakly closed *-derivation δ on N, with $D(\delta) = R(N)$, such that

$$(1 - \delta)^{-1} = R|_N ,$$

and hence

$$R = (1 - \delta)^{-1}E$$

This ends the proof of Theorem 5.1.

6. Alternative forms of the $O(t^{-1/2})$ estimate.

(The results in this section originated from a discussion with E. Brian Davies).

Adopt the basic assumptions and the notation from the beginning of the Introduction. In this section we will show that the basic estimate

$$\|\delta_k S_t\| \le Ct^{-1/2} \tag{1}$$

can be given alternative, and seemingly more tractable, forms provided it is known a priori that S is a holomorphic semigroup. (This is known a posteriori if g integrates). It turns out that the precise form of the alternative estimates depends on whether (1) is known only for small positive t, or for all positive t. If $G = \mathbf{R}^d$ and g integrates to an isometric representation, then using the explicit form of the heat kernel it is easy to establish estimate (1) for all $t > 0$ in this case. It may well be true that (1) is true for all positive t even for isometric representations of general Lie groups G, but this does not follow from the estimates in [BGJR1].

We consider the two cases of small positive t and all positive t separately.

Proposition 6.1. *Let* S *be a holomorphic semigroup on a Banach space* $\mathbf{B}$ *with generator* Δ, $\mathbf{D}$ *be a core for* Δ *with* $S_t \mathbf{B} \subseteq \mathbf{D}$ *for all* $t > 0$, *and* $H : \mathbf{D} \to \mathbf{B}$ *be linear and* *closable. The following conditions are equivalent.*

(1) *There is a constant* C_1 *such that*

$$\|HS_t\| \leq \frac{C_1}{t^{1/2}} \quad \textit{for} \quad 0 < t \leq 1 .$$

(2) *There is a constant* C_2 *such that*

$$\|Hx\| \leq C_2(\varepsilon\|\Delta x\| + \frac{\|x\|}{\varepsilon}) \quad \textit{for} \quad x \in \mathbf{D}, \quad 0 < \varepsilon \leq 1 .$$

(3) *There is a constant* C_3 *such that*

$$\|Hx\|^2 \leq C_3\|x\| \max(\|\Delta x\|, \|x\|) \quad \textit{for} \quad x \in \mathbf{D} .$$

Proof.

(2) => (3) If $0 < \|x\| \leq \|\Delta x\|$, put $\varepsilon = \left(\frac{\|x\|}{\|\Delta x\|}\right)^{1/2}$, giving

$$\|Hx\| \leq 2C_2\|x\|^{1/2}\,\|\Delta x\|^{1/2} = [4C_2^2\|x\| \max(\|\Delta x\|, \|x\|)]^{1/2} .$$

If $\|x\| > \|\Delta x\|$, put $\varepsilon = 1$, giving

$$\|Hx\| \leq C_2(\|\Delta x\| + \|x\|) < 2C_2\|x\|$$

$$= [4C_2^2\|x\| \max(\|\Delta x\|, \|x\|)]^{1/2} .$$

Thus (3) holds with $C_3 = 4C_2^2$.

(3) => (1) It is known that there is a constant K such that

$$\|\Delta S_t\| \leq \frac{K}{t} \quad \text{for} \quad 0 < t \leq 1$$

(see [Paz 1, p. 62] - rescaling of S may be needed if S is not bounded on a sector). Now

$$\|HS_tx\|^2 \leq C_3\|S_tx\| \max(\|\Delta S_tx\|, \|S_tx\|)$$

$$\leq C_3\, M\|x\| \max(\frac{K}{t}\|x\|, M\|x\|)$$

$$\leq C_3M\,\frac{\max(K,M)}{t}\,\|x\|^2 \quad \text{for} \quad 0 < t \leq 1 ,$$

where $M = \sup_{0 \leq t \leq 1} \|S_t\|$. Thus (1) is valid with $C_1 = \left(C_3M \max(K,M)\right)^{1/2}$.

(1) => (2) Choose M_1, $\omega \geq 0$ such that $\|S_t\| \leq M_1\, e^{\omega t}$.

Choose ε such that $0 < \varepsilon \leq 1$, and $\varepsilon^2\omega \leq 1/2$. Then

$$(1+\varepsilon^2\Delta)^{-1}y = \int_0^\infty dt\, e^{-t}\, S_{\varepsilon^2 t}y \quad \text{for} \quad y \in \mathbf{B} ,$$

$$\|HS_{\varepsilon^2 t}\| \le \frac{C_1}{\varepsilon t^{1/2}} \quad \text{for} \quad 0 < t \le 1 ,$$

$$\|HS_{\varepsilon^2 t}\| \le \|HS_{\varepsilon^2}\| \, \|S_{\varepsilon^2(t-1)}\|$$

$$\le \frac{C_1}{\varepsilon} M_1 e^{\omega\varepsilon^2(t-1)} \quad \text{for} \quad t > 1 .$$

Hence $T_\varepsilon = \int_0^\infty dt\, e^{-t} HS_{\varepsilon^2 t}$ exists, and

$$\|T_\varepsilon\| \le \int_0^1 dt \frac{C_1}{\varepsilon t^{1/2}} + \int_1^\infty dt \frac{C_1 M_1}{\varepsilon} e^{(\omega\varepsilon^2 - 1)t} e^{-\omega\varepsilon^2}$$

$$\le \frac{2C_1}{\varepsilon} + \frac{C_1 M_1}{\varepsilon(1-\omega\varepsilon^2)}$$

$$\le \frac{2(1+M_1)C_1}{\varepsilon} \quad \text{for} \quad 0 < \varepsilon \le 1, \ \varepsilon^2\omega < ½ .$$

Since H is closable,

$$\overline{H}(1+\varepsilon^2\Delta)^{-1} y = T_\varepsilon y \quad \text{for} \quad y \in \mathbf{B} .$$

For x in $\mathbf{D}$, putting $y = (1+\varepsilon^2\Delta)x$,

$$\|Hx\| = \|T_\varepsilon(1+\varepsilon^2\Delta)x\|$$

$$\le \frac{2(1+M_1)C_1}{\varepsilon} \|x + \varepsilon^2\Delta x\|$$

$$\le 2(1+M_1)C_1 \left(\varepsilon\|\Delta x\| + \frac{\|x\|}{\varepsilon}\right) \quad \text{for} \quad 0 < \varepsilon \le \min\left(1, \frac{1}{(2\omega)^{1/2}}\right).$$

Thus, if $\omega \le ½$, (2) is valid with $C_2 = 2(1+M_1)C_1$.

If $\omega > ½$, $0 < \varepsilon \le 1$, then

$$\varepsilon\|\Delta x\| + \frac{\|x\|}{\varepsilon} \le (2\omega)^{1/2} \sup_{0<\varepsilon' \le \frac{1}{(2\omega)^{1/2}}} \left(\varepsilon'\|\Delta x\| + \frac{\|x\|}{\varepsilon'}\right)$$

So (2) holds with $C_2 = 2(1+M_1)\, C_1(2\omega)^{1/2}$.

Proposition 6.2. *Let S be a bounded holomorphic semigroup on a Banach space $\mathbf{B}$ with generator Δ (so S is bounded on a sector), $\mathbf{D}$ be a core for Δ such that $S_t\mathbf{B} \subseteq \mathbf{D}$ for all $t > 0$, and $H : \mathbf{D} \to \mathbf{B}$ be linear and closable. The following conditions are equivalent:*

(1) *There is a constant C_1 such that*

$$\|HS_t\| \le \frac{C_1}{t^{1/2}} \quad \textit{for all} \quad t > 0 .$$

(2) *There is a constant* C_2 *such that*

$$\|Hx\| \le C_2\left(\varepsilon\|\Delta x\| + \frac{\|x\|}{\varepsilon}\right) \quad for \quad x \in \mathbf{D},\ \varepsilon > 0\,.$$

(3) *There is a constant* C_3 *such that*

$$\|Hx\|^2 \le C_3\|x\|\,\|\Delta x\| \quad for \quad x \in \mathbf{D}\,.$$

Proof.

(2) => (3). Put $\varepsilon = \left(\frac{\|x\|}{\|\Delta x\|}\right)^{1/2}$, $C_3 = 4C_2^2$.

(3) => (1). There exists K such that $\|\Delta S_t\| \le \frac{K}{t}$ for all $t > 0$. Hence

$$\begin{aligned}\|HS_tx\|^2 &\le C_3\,\|S_tx\|\,\|\Delta S_tx\| \\ &\le C_3\,M\frac{K}{t}\,\|x\|^2\,.\end{aligned}$$

Thus (1) is valid with $C_1 = (C_3\,MK)^{1/2}$.

(1) => (2). For any $\varepsilon > 0$,

$$\begin{aligned}\|\overline{H}(1+\varepsilon^2\Delta)^{-1}\| &= \left\|\int_0^\infty dt\, e^{-t}\, HS_{\varepsilon^2 t}\right\| \\ &\le \int_0^\infty dt\, e^{-t}\,\frac{C_1}{\varepsilon t^{1/2}} \\ &= \frac{C_2}{\varepsilon}, \quad \text{where} \quad C_2 = C_1\int_0^\infty dt\,\frac{e^{-t}}{t^{1/2}}\,.\end{aligned}$$

Hence

$$\|Hx\| \le \frac{C_2}{\varepsilon}\|(1+\varepsilon^2\Delta)x\| \le C_2\left(\varepsilon\|\Delta x\| + \frac{\|x\|}{\varepsilon}\right).$$

Acknowledgements

We have profited from discussions with Man-Duen Choi, E. Brian Davies, George A. Elliott, David E. Evans, Palle E.T. Jørgensen and Jun Tomiyama. Most of this work was done while the three authors visited the Mathematics Research Centre, University of Warwick with support from SERC. O.B. had a travel grant from NAVF.

References

[AB1] Arendt, W. and C.J.K. Batty, Tauberian theorems and stability of one-parameter semigroups, preprint (1987).

[AOR1] Allan, G.R., A.G. O'Farrell and T.J. Ransford, A Tauberian theorem arising in operator theory, preprint (1986).

[BGJR1] Bratteli, O., F.M. Goodman, P.E.T. Jørgensen and D.W. Robinson, The heat semigroup and integrability of Lie algebras, submitted to *J. Functional Anal.*

[BR1] Bratteli, O. and D.W. Robinson, Operator Algebras and Quantum Statistical Mechanics I, Second Edition, Springer Verlag, New York (1987).

[EMOT1] Erdélyi, A., W. Magnus, F. Oberhettinger and F.G. Tricomi, Higher Transcendental Functions, Volume II, McGraw-Hill, New York-Toronto-London (1953).

[Her 1] Herman, R.H., Unbounded derivations, *J. Functional Anal.* **20** (1975), 234-239.

[JM1] Jørgensen, P.E.T. and R.T. Moore, Operator Commutation Relations, Reidel, Boston-Dordrecht (1984).

[KT1] Katznelson, Y. and L. Tzafriri, On power bounded operators, *J. Functional Anal.* **68** (1986), 313-328.

[LM1] de Leeuw, K. and H. Mirkil, Majorations dans L_∞ des opérateurs différentiels à coefficients constants, *C. R. Acad. Sci. Paris* **254** (1962), 2286-2288.

[LM2] de Leeuw, K. and H. Mirkil, A priori estimates for differential operators in L^∞-norm, *Illinois J. Math.* **8** (1964), 112-124.

[Nel 1] Nelson, E., Analytic Vectors, *Ann. Math.* **70** (1959), 572-615.

[Paz 1] Pazy, A., Semigroups of Linear Operators and applications to Partial Differential Equations, Springer-Verlag, New York (1983).

[RN1] Riesz, F. and B. Sz-Nagy, Leçons d'analyse fonctionnelle, 6th ed., Gauthier-Villars, Paris (1972).

[Rob 1] Robinson, D.W., The differential and integral structure of Lie groups, Canberra preprint (1987).

[Rot 1] Rota, G.-C., Reynolds operators, *Proc. Symp. Appl. Math.* **16** (1964), 70-83.

[Tak 1] Takesaki, M., Conditional expectations in von Neumann algebras, *J. Functional Analysis* **9** (1972), 306-321.

[Tom 1] Tomiyama, J., On the projection of norm 1 in W*-algebras I, II, III, *Proc. Jap. Acad. Sci.* **33** (1957), 608-612; *Tohoku Math. J.* **10** (1958), 204-209; ibid. **11** (1959), 125-129.

[Yos 1] Yosida, K., Functional Analysis, Second Edition, Springer-Verlag, Berlin-Heidelberg-New York (1968).

C*ALGEBRAS IN SOLID STATE PHYSICS
2D Electrons in a uniform magnetic field[&]

Jean BELLISSARD [#]

Centre de Physique Théorique [§]
CNRS- Luminy, Case 907
F-13288, MARSEILLE, CEDEX 09
(FRANCE)

[&] *Talk given at the Warwick Conference on Operator Algebras (July 1987)*

Abstract: *Technics recently developped in non commutative geometry through properties of C^*Algebras are presented without proofs here to investigate some properties of 2D electrons in a uniform magnetic field. The Peierls substitution, a commonly used approximation for Bloch electrons, is justified and leads to the Rotation Algebra. A new differential calculus, analogous to the Ito calculus for stochastic processes, is introduced to investigate the fine structure of the energy spectrum. We announce the proof of the Wilkinson Rammal formula according to which the derivative of the energy gap boundaries are discontinuous at each rational values of the magnetic flux. We also announce that the derivative is continuous at irrational values of this flux. At last we review and improve the results previously obtained for the Quantum Hall Effect and sketch the proof that in the region of localized states the Hall conductance exhibits plateaux at integer values of the universal constant e^2/h.*

September 1987
CPT-87/P.2047

Université de Provence, Marseille
§ Laboratoire Propre, LP-7061, Centre National de la Recherche Scientifique.

C*ALGEBRAS IN SOLID STATE PHYSICS: 2D Electrons in a Uniform Magnetic Field.

Jean BELLISSARD
Centre de Physique Théorique and Université de Provence,
CNRS, Luminy, Case 907, 13288, Marseille, Cedex 09, France.

Abstract : Technics recently developped in non commutative geometry through properties of C*Algebras are presented without proofs here to investigate some properties of 2D electrons in a uniform magnetic field. The Peierls substitution, a commonly used approximation for Bloch electrons, is justified and leads to the Rotation Algebra. A new differential calculus, analogous to the Ito calculus for stochastic processes, is introduced to investigate the fine structure of the energy spectrum. We announce the proof of the Wilkinson Rammal formula according to which the derivative of the energy gap boundaries are discontinuous at each rational value of the magnetic flux. We also announce that the derivative is continuous at irrational values of this flux. At last we review and improve the results previously obtained for the Quantum Hall Effect and sketch the proof that in the region of localized states the Hall conductance exhibits plateaux at integer values of the universal constant e^2/h.

I- Introduction :

Several problems in Solid State Physics proved recently to be understandable through using technics developped in non commutative topology and geometry. Let us mention the study of Schrödinger operators with almost periodic potential [14, 86] which may describe the behavior of a Bloch electron in a uniform magnetic field [see 88 for a review and discussion below], the stability of a dynamical system near a quasi periodic orbit [69, 86], the periodic or quasi periodic solutions of the KdV equation [35, 69, see 86 and references therein], the ground state properties of one dimensional organic conductors [7, 8], the metal-superconductor for superconductors in a network [3, 40, 72, 80], weak localization in normal metal networks [32, 33, 34] or the electronic properties of quasi crystals [36, 58, 66, 83]. These models are described by operators which are not translation invariant but are homogeneous under space translations. A natural observable algebra is generated by the operators obtained by translating the hamiltonian in space. In most of these models the energy spectrum is nowhere dense [15, 69, see 86 for a review], and a natural gap labelling is provided by the K_0 group of the C*algebra [16].

The behaviour of a Bloch electron in a magnetic field is known to be one of the most difficult and challenging problem in Solid State Physics. The first important theoretical contributions go back to the thirties whith the works of Landau [63] and Peierls[76], and since this time, there has been no expert in the field who did not contribute to its study. A rough estimate shows that at least fifty articles of major importance were published between 1950 and 1965 on the sole question of how to compute the band spectrum in that case [In this work we have selected only 2, 12, 21, 23, 24, 25, 26, 42, 49, 53, 64, 67, 70, 76, 78, 105]. The reason is that magnetic fields break the parity or the time reversal symetries, and provide then a very efficient and technically simple tool in investigating microscopic properties of a metal. One of the oldest example of such use is the Hall effect discovered in 1880 by Hall [46] : through a very simple experiment it is possible to check whether the charge carriers of a current are electrons or holes. Measurements of the Fermi surfaces involve magnetic fields as well [68], and the systematic measurement of the magneto-resistance give many informations as it has been illustrated in the more recent study of weak localization effects in disordered systems [19].

Since 1980, two sets of experiments provided a spectacular contribution to this knowledge: the discovery of the quantum Hall effect in 1980, by Von Klitzing, Pepper and Dorda [57, see also 4, 97], and the sequence of experiments by Chaussy, Pannetier and Rammal and their various collaborators [34, 39, 43, 72, 73, see also 13] on arrays of super conductors or dirty metals (we will speak of the "Grenoble experiments" throughout this paper). They will be emphasized here because they are amazingly accurate and provide a strong motivation for a mathematical physicist to go as far as possible in understanding the mathematical framework of the models describing them, but also because C*Algebras and their non commutative geometry build up a natural framework which will prove very efficient in giving exact formulæ and accurate estimates as well.

This report will contain four sections. Section II concerns spectral properties of a Schrödinger operator describing a Bloch electron in a magnetic field. We review the results obtained during the fifties and the sixties. We have updated them by adding recent or new mathematical results, in particular concerning the Peierls substitution. Section III contains a description of the Rotation Algebra. We have included a new algebraic object, a sort of a "Ito" differential calculus [see also 54] which helps in deriving several formulæ like the "Streda" formula [89] and the Wilkinson Rammal one in the next paragraph. Section IV is devoted to new results obtained during the last year concerning the properties of the Hofstadter like spectra. They have leaded to recent experiments by the Grenoble group [39, 99, 101]. With the help of the semi classical approach of Briet, Combes and Duclos [22, 27] and of Hellfer, Sjöstrand [50, 51], we indicate how to derive the "Wilkinson-Rammal" formula for the derivative of the boundary of the spectrum [101,103]. Section V concerns the Quantum Hall Effect. This subject has been already presented in a conference at Bad Schandau [18], and we will emphasize here only the part of the argument concerning the Kubo formula, its connection with Chern classes, the calculation of the integers involved in the quantization of the Hall conductance using the Connes formula, and the connection with the finiteness of the localization length. Indeed, since the Bad Schandau conference this latter result has been understood in a better and more natural way in term of non commutative Sobolev norms which actually also appeared independently in a recent work of A. Connes on non commutative riemanian manifolds [30]. The reader is invited to look at the Bad Schandau paper if he is interested in the physical background and the need for using non commutative geometry.

Acknowledgements : *This work has benefitted of many opportunities provided by various institutions which invited me during the last five years. The work on the Quantum Hall Effect started while I was at Princeton University in 1983, and the final version was settled at Caltech during the spring of 1986. The work concerning Bloch electrons in a magnetic field was partly solved in 1985 while in Marseilles and the most recent results were achieved at the RIMS in Kyoto in February, at the IHES during May, and also while visiting the Warwick University in July of 1987. I want to thank these institutions and especially Prs. E. Lieb, A. Wightman, B Simon, H. Araki, A. Connes and D. Evans for confidence, financial support, and/or scientific exchanges. I am also indebted to many mathematicians and physicists who gave me with patience and entousiathm an enormous amount of informations and ideas put together in this report. I will refer to them in the core of the text and I thank them for their contributions.*

II- 2D Bloch electrons in a magnetic field :

In the simplest approximation, electrons in a metal are usually described as a free Fermi gas [see 68]. Each electron is independent from the others, and only the statistics created by the Pauli principle gives rise to a collective behavior. The next approximation consists in taking into account the periodic structure of the metal, which implies understanding the cristallographic structure of the atomic arrangement. The main result in taking these effects into account is in adding to the kinetic energy of each electron a potential energy term, local and periodic in space, describing the external forces created by the ions on the electrons. But still, the electrons depend upon each others only through the statistics in this approximation, and the ions are sufficiently heavy to be represented as fixed. The Bloch theory, an extension for 2D or 3D quantum physics of the Floquet theory for periodically driven dynamical systems, gives rise to a band spectrum with an absolutely continuous spectral measure called the density of state. The Fermi Dirac statistics leads to the notion of Fermi energy which coincides at low temperature with the chemical potential. Because the corresponding Fermi temperatures are usally quite big (between 5 and 16×10^4 K), the zero temperature approximation is almost always very good even at the room temperature. For these reasons all the difficulties are concentrated only on the spectral properties of the one electron hamiltonian. In this work, we will ignore the many body effects which only occur as small corrections in most cases, unless the conditions of the experiments are to emphasise collective effects due to the electron-electron Coulomb interactions, as it occurs in the Fractional Quantum Hall Effect and probably in the theory of High Temperature Superconductors.

Eventhough real samples have three dimensional homogeneity properties, many modern devices where magnetic properties are emphasized have been shown to be mainly two dimensional. This is definitely the case both for the Quantum Hall Effect and for the Grenoble experiments. Hence we will restrict ourselves to the 2D case.

The conventional way in describing the one electron hamiltonian for a Bloch electron in a magnetic field is to make the substitution $\mathbf{P} \to \mathbf{K} = \mathbf{P} - e\mathbf{A}$, where $\mathbf{P}$ is the momentum operator and $\mathbf{A}=(A_1,A_2)$ the vector potential, in the Schrödinger operator. In 2D this leads to the Hamiltonian :

$$H = \frac{1}{2m} \{(P_1 - eA_1)^2 + (P_2 - eA_2)^2\} + V(Q_1,Q_2) \qquad (1)$$

where $\mathbf{Q} = (Q_1,Q_2)$ is the position operator, V is periodic in $\mathbf{Q}$ with period group Γ which we will call by extension the Bravais lattice [see 68 for details], m is the electron mass, e the electron charge and, if B is the magnetic field in the direction perpendicular to the sample, $\mathbf{A}$ satisfies :

$$\partial_1 A_2 - \partial_2 A_1 = B \tag{2}$$

Actually, one can also describe the energy operator through a lattice approximation leading to various discrete models in particular if one wants to use strange lattices as the Penrose lattice [13] or the "Serpinsky gasket" [43], which is a first step toward the understanding of electronic properties of percolation clusters of superconductors [3, 40, 80]. In the present section we will see how these approximations are justified. However it is important to notice that the algebraic framework will apply whatever the explicit form of H.

II-1) Strong field expansion for Landau's band-

For V = 0 the spectrum of H has been performed first by Landau in 1930 [63]. It comes from the remark that $\mathbf{K}$ obey to the canonical commutation relations namely :

$$[K_1,K_2] = iehB/2\pi \tag{3}$$

Therefore $H = H_L$ becomes the energy operator for an harmonic oscillator. The spectrum is then given by eigenvalues :

$$E_n = (2n+1)E_0 \qquad E_0 = h/2\pi\ (eB/m) \qquad n = 0,1,2,\ldots \tag{4}$$

These eigenvalues have actually an infinite degeneracy. Adding V will broaden these "Landau levels". We will restrict ourselves to the case for which V is bounded (one could extend the study to the case for which V is relatively bounded with respect to H_L). If the bound is small enough, the resulting spectrum will be made of well separate Landau bands. Thanks to (4), the separation will increase with the magnetic field in such a way that the separate band picture will describe the strong field approximation.

Let Π_n be the nth eigenprojection of H_L . The spectrum of H in the vicinity of the nth band $\mathcal{B}_n$ is described through the so-called "projection method" [38], leading to the effective one band hamiltonian :

$$H_n(z) = \Pi_n H \Pi_n + \Pi_n H (1-\Pi_n) \frac{1}{z1-(1-\Pi_n)H(1-\Pi_n)} (1-\Pi_n)H\Pi_n \tag{5}$$

which is defined and analytic with respect to z in a neighbourhood of the nth band. Then, z belongs to $\mathcal{B}_n$ if and only if z belongs to the spectrum of $H_n(z)$, an implicit equation for z.

In order to analyze this effective hamiltonian, let us introduce a new set of canonical variables, besides $\mathbf{K}$. We first choose the symetric gauge namely the solution of (2) given by :

$$A_\mu(Q) = -B/2\, \varepsilon_{\mu\nu} Q_\nu \qquad \varepsilon_{12} = 1 = -\varepsilon_{21} \qquad \varepsilon_{11} = 0 = \varepsilon_{22} \tag{6}$$

If we set :

$$k_\mu = (1/2)Q_\mu + (1/eB)\, \varepsilon_{\mu\nu} P_\nu \tag{7}$$

we get :

$$[K_\mu,k_\nu] = 0 \qquad [k_1, k_2] = -ih/(2\pi eB) \tag{8}$$

and:

$$Q_\mu = k_\mu - (1/eB)\varepsilon_{\mu\nu} K_\nu \tag{9}$$

Then we expand V in Fourier series, and we use the well known formulæ giving the matrix elements of an operator in the eigenstates of the harmonic oscillator, to get the following result [26i, 107]:

Theorem 1 : 1) the nth band effective hamiltonian $H_n(z)$ belongs to the C*Algebra generated by the unitary operators

$$U = e^{i\langle b_1|k\rangle} \qquad V = e^{i\langle b_2|k\rangle} \tag{10}$$

where $\mathbf{b}_1,\mathbf{b}_2$ are the generators of the reciprocal lattice Γ^*. They satisfy :

$$U\,V = e^{2i\pi(1/\alpha)}\,V\,U \qquad \alpha = BS/(h/e) \tag{11}$$

where S is the area of the unit cell of the Bravais lattice Γ.
2) If V is continuous on the plane and satisfies $|V| < E_0$, $H_n(z)$ admits an expansion converging uniformly in a complex neighbourhood of the nth band, of the form :

$$H_n(z) = \Pi_n\{(2n+1)E_0 + \sum_{b\in\Gamma^*} \underline{V}(\mathbf{b})\, J_n(\mathbf{b};\alpha)\, W(\mathbf{b}) + O(|V|^2/E_0)\} \qquad (12)$$

where the $\underline{V}(\mathbf{b})$'s are Fourier coefficients of V, $J_n(\mathbf{b};\alpha)$ a complex number with modulus less than one, and :

$$W(m_1\mathbf{b}_1 + m_2\mathbf{b}_2) = e^{-i\pi(1/\alpha)m_1 m_2}\, U^{m_1}\, V^{m_2} \qquad (13)$$

◊

Remark : Equation 11 shows that α represents the ratio between the magnetic flux through the unit cell of the Bravais lattice and the flux quantum h/e. In the case of a superconductor h/e must be replaced by h/2e since the charge carriers are Cooper pairs rather than electrons [41].

II-2) Weak field expansion : the Peierls substitution-

The previous analysis is worthless in most of the usual applications when the magnetic field is hardly strong enough to justify the Landau bands picture. At low field however one should see a pattern close in some sense to the Bloch energy band picture. In 1933, Peierls [76] proposed the following rule to get an effective band hamiltonian in presence of a weak magnetic field: let $E_n(\mathbf{k})$ be the nth energy function of the Schrödinger operator whithout magnetic field. It is defined on the Brillouin zone, and periodic in the quasi momentum $\mathbf{k}$ with the reciprocal lattice as period group. The *Peierls substitution* consists in replacing the vector $\mathbf{k}$ by the vectorial operator $\mathbf{K} = \mathbf{P} - e\mathbf{A}$. To avoid arbitrariness in this procedure due to the non commutativity of K_1 and K_2, let us agree with the Weyl quantization rule : we expand E_n in Fourier series in $\mathbf{k}$ and we replace the trigonometric monomial $e^{i\langle a|k\rangle}$ where $\mathbf{a}$ belongs to the Bravais lattice Γ by the operator $e^{i\langle a|K\rangle}$. In this way, if E_n is smooth enough, we get a bounded self adjoint operator H_n belonging to the C*Algebra generated by the $e^{i\langle a|K\rangle}$'s, namely by two unitaries U and V defined by :

$$U = e^{i\langle \mathbf{a}_1|K\rangle} \qquad V = e^{i\langle \mathbf{a}_2|K\rangle} \qquad U\,V = e^{2i\pi\alpha}\, V\,U \qquad (14)$$

where $\mathbf{a}_1$ and $\mathbf{a}_2$ generate Γ, and α is defined by (11).

If we accept this rule for correct description of the effective band hamiltonian at weak field, we get a description very similar to the strong field one with the difference that now $1/\alpha$ is replaced by α itself. In this sense can we speak of a *strong field - weak field duality* [64] which does not seem to have been exploited unless perhaps in a paper by Azbel' [12] in 1964 to justify a qualitative description of the band spectrum at non zero field.

The justification of the Peierls rule together with the construction of a systematic weak field expansion has been the subject of many papers during the fifties. Following an idea of Slater, Luttinger in 1951[67] used Wannier functions to derive the Peierls rule as the lowest order approximation. In 1952, Onsager [70] and Adams II [2] used a semi classical approach to understand the properties of the corresponding hamiltonian. These two ingredients are actually essential in the analysis of the energy spectrum. The first one will give a power law expansion namely the perturbation theory, the second one will give exponentially small corrections in the magnetic field, coming from tunneling in phase space. The number of papers contributing to this program is too high for such a short report here and we will skip it, hopping to be able to publish it later in a more complete version [107] .

The first ingredient in justifying the Peierls substitution is the following result due to Avron and Simon [9]. Let Σ be the spectrum of H. Σ is a closed subset of the real line and therefore its complement is the union of a countable family of open intervals (a_i, b_i) called the gaps.

Theorem 2 : If H is the Schrödinger operator given by (1), the boundaries of the gaps of its spectrum vary continuously with the magnetic field at eeach points where they are open.

◊

A similar theorem was previously proved by G.Elliott [37] for a self adjoint operator in the rotation algebra (cf theorem 5 below). Actually the Elliott result can be extend in a much more general situation (theorem12 below) by using a different argument including all kind of models encountered in physics without being restricted to the Schrödinger case.

From this result it follows that if H admits some band spectrum at zero magnetic field, the corresponding gaps will survive for a while as the magnetic field is turned on.

At zero field the Bloch theory applies, namely H can be decomposed through Bloch waves as follows : each band of multiplicity one is described by a function $\phi(x;k)$ such that

$$\phi(\mathbf{x}+\mathbf{a};\mathbf{k}) = e^{2i\pi \mathbf{k}\mathbf{a}/h}\,\phi(\mathbf{x};\mathbf{k}) \qquad \phi(\mathbf{x};\mathbf{k}+\mathbf{b}) = \phi(\mathbf{x};\mathbf{k}) \tag{15}$$

for every **a** in the Bravais lattice and **b** in the reciprocal lattice. Moreover if H(k) denotes the self adjoint operator obtained by taking the partial differential operator (1) with B=0 and the Bloch boundary conditions (15), $\phi(x;k)$ satisfies :

$$\{H(\mathbf{k})\,\phi\}(\mathbf{x};\mathbf{k}) = E(\mathbf{k})\,\phi(\mathbf{x};\mathbf{k}) \qquad \text{with} \qquad E(\mathbf{k}+\mathbf{b}) = E(\mathbf{k}) \tag{16}$$

where E is the energy band function. One then introduces the Wannier function $\psi(x)$ as :

$$\psi(\mathbf{x}) = \int_B d^2\mathbf{k}\;\phi(\mathbf{x};\mathbf{k}) \tag{17}$$

where B is the Brillouin zone, namely the unit cell of the reciprocal lattice. One can show that if V is relatively bounded with respect to the laplacian in 2D, the energy band function is analytic in **k** [102], and that because H is real, the Bloch wave is also analytic in **k** (the case of bands with multiplicity is more involved and we will avoid it here). Then $\psi(x)$ is exponentially decreasing away from zero and if $\psi_a(x) = \psi(x-a)$ (with a in Γ) we get an orthonormal basis [see for instance 68]:

$$\int_{R^2} d^2\mathbf{x}\;\psi_{\mathbf{a}}(\mathbf{x})\,\psi_{\mathbf{a}'}(\mathbf{x}) = \delta_{\mathbf{a},\mathbf{a}'} \tag{18}$$

Following Luttinger [67], we introduce the modified Wannier functions by mean of the magnetic translation [105] :

$$\psi_{\mathbf{a},B}(\mathbf{x}) = e^{2i\pi eB/h\,(\mathbf{a}\wedge\mathbf{x})}\,\psi_{\mathbf{a}}(\mathbf{x}) \tag{19}$$

If the field is weak, this family is almost orthonormal in the sense that inner product of two of them will decay exponentially fast in the difference **a–a'**. We then introduce the orthogonal projection Π onto the subspace generated by the $\psi_{a,B}$'s. It becomes norm continuous in B and converges to the band projection as B->0. As before we define by mean of the projection method the following effective hamiltonian :

$$H_{eff}(B,z) = \Pi H\Pi + \Pi H(1-\Pi)\,\frac{1}{z1-(1-\Pi)H(1-\Pi)}\,(1-\Pi)H\Pi \tag{20}$$

This is a norm analytic function of z in a complex neighbourhood of the band, and z belongs to the part of the spectrum of H in this band if and only if z belongs to the spectrum of $H_{eff}(B,z)$, again an implicit equation for z.

At last we introduce the orthonormal basis $\eta_{a,B}$ obtained from the $\psi_{a,B}$'s by diagonalising the matrix of inner products. Then every element ψ of the subspace generated by Π can be identified with the sequence $\{f(a);a\in\Gamma\}$ of its coordinates which is square summable. Thanks to this description one gets [107]:

<u>Theorem 3</u> : The band hamiltonian $H_{eff}(B,z)$ given by (20) can be expanded as :

$$H_{eff}(B,z) = \sum_{\mathbf{a}'\in\Gamma} E_{eff}(z;\mathbf{a}')\,W(\mathbf{a}') \qquad \text{with} \qquad \{W(\mathbf{a}')f\}(\mathbf{a}) = e^{i\pi eB/h(\mathbf{a}\times\mathbf{a}')}\,f(\mathbf{a}+\mathbf{a}') \tag{21}$$

where the W(a)'s obey the Weyl commutation relations. Moreover, the coefficients $E_{eff}(z;a)$ of the expansion satisfy :

$$E_{eff}(z;\mathbf{a}) = \underline{E}(\mathbf{a}) + O(|B|) \tag{22}$$

where the $\underline{E}$ (a)'s are the Fourier coefficients of the energy band function E(k).
If a_1 and a_2 generate Γ, one has :

$$W(m_1\mathbf{a}_1+m_2\mathbf{a}_2) = e^{-i\pi\alpha m_1 m_2}\,U^{m_1}V^{m_2} \qquad \text{with} \qquad UV = e^{2i\pi\alpha}\,VU \tag{23}$$

◊

From (22) the energy band function E(k) gives rise to the effective hamiltonian after performing the Peierls substitution up to an error of order B and the projection method give an exact effective band hamiltonian.

III- The Rotation Algebra :

III-1) Topology–

In the previous section we have seen that every operator of interest can actually be expanded in term of trigonometric polynomials in two unitaries U and V such that :

$$U\,V = e^{2i\pi\alpha}\,V\,U \qquad (1)$$

The parameter α is the ratio of the magnetic flux through the unit cell of the Bravais lattice, and the flux quantum h/e (see §II eq(11)). In the strong field expansion however, we have seen that α must be replaced by $1/\alpha$. The C*algebra generated by two such unitaries is called the rotation algebra and will be denoted by $\mathcal{A}_\alpha$. It has been introduced and studied by M.Rieffel [82] in 1977. However, in the early seventies A.Grossman [44] showed that the Von Neumann algebra generated by them is type II when α is irrational. This result was a source of difficulties in understanding the spectral behaviour of Bloch electrons in a magnetic field. Actually it is now known that :

Theorem 4 : 1) $\mathcal{A}_\alpha = \mathcal{A}_{\alpha+1}$; $\mathcal{A}_\alpha$ and $\mathcal{A}_{\alpha'}$ are isomorphic if and only if $\alpha = \pm\,\alpha'$ (mod.1)

2) If α is irrational, $\mathcal{A}_\alpha$ is simple.

3) Let W_1 and W_2 two qxq unitary matrix such that :

$$W_1\,W_2 = e^{2i\pi p/q}\,W_2\,W_1 \qquad W_2{}^q = W_2{}^q = 1 \qquad (2)$$

then $\mathcal{A}_{p/q}$ is isomorphic to the set of matrix valued functions F(**k**) over $\mathbf{R}^2$ such that :

$$(W_1{}^{m_2}W_2{}^{-m_1})\,F(\mathbf{k})\,(W_1{}^{m_2}W_2{}^{-m_1})^* = F(\mathbf{k}+\mathbf{m}) \qquad \mathbf{m} = (m_1,m_2)\in\mathbf{Z}^2 \qquad (3)$$

◊

Remark : The periodicity in $1/\alpha$ for a strong field was known by the physicists from a long time: Landau noticed it in 1930 [63] but doubted if it where possible to observe it. He was not aware of the experiments of De Haas and Van Alven [68] at about the same time exhibiting such periodicity which remained mysterious quite a long time. It was not before the work of L.Onsager in 1952 [70] that a semi-classical theory was provided for the phenomena which proved to be one of the basic experimental method in the measurement of the Fermi surface.

Corollary : If α is irrational, every non zero representation of $\mathcal{A}_\alpha$ is faithful.

◊

A trace on $\mathcal{A}_\alpha$ is defined as the unique state τ such that :

$$\tau\,(U^{m_1}\,V^{m_2}) = 0 \qquad \text{if} \qquad \mathbf{m} = (m_1,m_2) \neq 0 \qquad (4)$$

Proposition 1 : For every α the trace on $\mathcal{A}_\alpha$ is faithful.

◊

Since α is proportional to the magnetic field, it is liable to vary continuously. The theorem 4 shows that $\mathcal{A}_\alpha$ does not vary continuously. However, as shown by G. Elliott the family $\{\mathcal{A}_\alpha;\ \alpha\in\mathbf{R}\}$ is a continuous field of C*Algebras [31,108]. A simple way to construct this field, consists in using the Universal Rotation Algebra $\mathcal{A}$ defined as the C*Algebra generated by three unitaries U,V,and λ, such that :

$$[U,\lambda] = [V,\lambda] = 0 \qquad\qquad U\,V = \lambda\,V\,U \qquad (5)$$

Then we define by ρ_α the canonical *homomorphism from $\mathcal{A}$ onto $\mathcal{A}_\alpha$ such that :

$$\rho_\alpha\,(U) = U \qquad \rho_\alpha\,(V) = V \qquad \rho_\alpha\,(\lambda) = e^{2i\pi\alpha}\,\mathbf{1} \qquad (6)$$

Elliott's result [37] is then equivalent to :

Theorem 5 : 1) Let A be an element of the Universal Rotation Algebra. Then the maps $\alpha \to \|\rho_\alpha(A)\|$ and $\alpha \to \tau\{\rho_\alpha(A)\}$ are continuous.

2) Let H be self adjoint in $\mathcal{A}$. Then, the gap boundaries of the spectrum of $\rho_\alpha(H)$ are continuous functions of α.

◊

III-2) Differential structure-

As remarked by A.Connes [28], one can see $\mathcal{A}_\alpha$ as a non commutative 2D torus by identifying U and V with the coordinate functions. This allowed him to define two derivations as follows both on $\mathcal{A}$ and $\mathcal{A}_\alpha$:

$$\partial_1 U = 2i\pi\, U \qquad \partial_1 V = 0 \qquad \partial_2 U = 0 \qquad \partial_2 V = 2i\pi\, V \tag{7}$$

These derivations commute and are actually the infinitesimal generators of a two parameter group of automorphism representing an action of the 2-torus defined by :

$$\eta_{x,y}(\, U^{m_1} V^{m_2}) = e^{2i\pi(m_1 x + m_2 y)}\, U^{m_1} V^{m_2} \tag{8}$$

$C^k(\mathcal{A}_\alpha)$ will denote the intersection of the domains of every powers of order at most k of these derivations. Thus A belongs to $C^k(\mathcal{A}_\alpha)$ means that the map $(x,y) \to \eta_{x,y}(A)$ is C^k in the norm topology. One also introduces the Sobolev space $\mathcal{H}_s$ as the Hilbert space obtained by completing $C^k(\mathcal{A}_\alpha)$ (for $k \geq s$) under the norm :

$$\| A \|_s = \{\tau(|(-\Delta/4\pi^2+1)^{s/2} A|^2)\}^{1/2} \quad \text{where} \quad \Delta = \partial_1{}^2 + \partial_2{}^2 \tag{9}$$

Using the Fourier expansion in terms of U and V, it is possible to define this norm for any non negative real value of s. This family of space possesses many of the properties of the usual Sobolev spaces.

III-3) The Chern class of a projection-

Let now P be a projection of $\mathcal{A}_\alpha$. If P belongs to $\mathcal{H}_1$, the Chern class of P is defined by analogy with the 2-torus as :

$$\mathrm{Ch}(P) = (1/2i\pi)\, \tau\, (P[\partial_1 P, \partial_2 P]\,) \tag{10}$$

Then one can check that Ch assumes the following properties [6, 29] :

i) Invariance : let P and Q be two equivalent projections in $\mathcal{H}_1$ namely such that $SS^* = P$ and $S^*S = Q$ for some S in $C^1(\mathcal{A}_\alpha)$, then Ch(P) = Ch(Q).
ii) Additivity : let P and Q be two projections in $\mathcal{H}_1$ such that PQ = QP = 0, then :
$\mathrm{Ch}(P \oplus Q) = \mathrm{Ch}(P) + \mathrm{Ch}(Q)$

The invariance property implies in particular that Ch extends to all projections of $\mathcal{A}_\alpha$ for any projection is equivalent to a C^1 projection [29]. Moreover it implies the homotopy invariance namely if s->P(s) is a norm continuous family of projection in $\mathcal{A}_\alpha$, Ch(P(s)) is constant for two projections close enough are equivalent [75]. The main result concerning the Chern class is the following :

Theorem 6 : If P is a projection of $\mathcal{A}_\alpha$, its Chern class is an integer. ◊

This result was proved in this particular case by A. Connes in 1980 [28], but later results of Connes allow to extend it in a much bigger generality in order to be used for the Quantum Hall Effect [18]. A sketch of the proof will be given at the end of this paper.

III-4) A Ito like derivation in $\mathcal{A}_\alpha$-

How do gaps boundaries of the spectrum of H vary whith the magnetic field ? This question motivates us to introduce in $\mathcal{A}$ the derivation with respect to α. More precisely let A be polynomial in U,V and λ. One can expand A as follows:

$$\rho_\alpha(A) = \sum_{m \in Z} a(m;\alpha)\, W(m) \qquad \text{with} \qquad W(m_1,m_2) = e^{i\pi\alpha m_1 m_2}\, U^{m_1} V^{-m_2} \tag{11}$$

We define the operation ∂ by the following formula :

$$\rho_\alpha(\partial A) = \sum_{m \in Z^2} \frac{\partial a(m;\alpha)}{\partial \alpha} W(m) \quad (12)$$

$C^1(\mathcal{A})$ will denote the completion of the set of polynomial under the norm :

$$\|A\|_{C^1} = \|\partial A\| + \|\partial_1 A\| + \|\partial_2 A\| \quad (13)$$

This operation satisfies the following rules [107]:

Theorem 7 : 1) If A and B belong to $C^1(\mathcal{A})$:

$$\partial(AB) = \partial A\, B + A\partial B + i/4\pi\, \{\partial_1 A \partial_2 B - \partial_2 A \partial_1 B\} \quad (14)$$

2) If A belongs to $C^1(\mathcal{A})$ and if it is invertible, its inverse belongs to $C^1(\mathcal{A})$ and

$$\partial(A^{-1}) = -A^{-1}\{\partial A + i/4\pi(\partial_1 A\, A^{-1}\partial_2 A - \partial_2 A\, A^{-1}\partial_1 A)\}\, A^{-1} \quad (15)$$

3) Streda's formula : if $P \in C^1(\mathcal{A})$ is such that for α in the interval (a,b), $\rho_\alpha(P)$ is a projection, then one gets in (a,b) :

$$\frac{\partial}{\partial \alpha} \tau(\rho_\alpha(P)) = Ch(\rho_\alpha(P)) \quad (16)$$

◊

Corollary : The C*algebra $\mathcal{A}$ has no non trivial projection.

◊

The proof of 1) and 2) in theorem 7 is a simple matter of calculation, and to prove 3) it is enough to remark that a projection satisfies $P^2 = P$, that $\tau(AB) = \tau(BA)$, since τ is a trace and that $\partial\tau(\rho_\alpha(A))/\partial\alpha = \tau(\rho_\alpha(\partial A))$.

The proof of the corollary starts with the Streda formula. If P belongs to $C^1(\mathcal{A})$, by theorem 6 the right hand side of (16) is a fixed integer for all α. Since the left hand side is the derivative of a periodic function of α, we must get $Ch(\rho_\alpha(P)) = 0$. We then integrate both members of (16) from zero to one and we get a formula which extends to any projection of $\mathcal{A}$. Now, from Pimsner and Voiculescu [77], we know that $\tau(\rho_\alpha(P)) = m+n\,\alpha$, and from the Streda formula it follows that $n = 0$. Since P is a projection and $\tau(\rho_\alpha(P))$ is continuous in α, we get $m = 0$ or 1. The trace being faithfull (prop. 1), it follows that $\rho_\alpha(P)$ is either **0** for all α's or **1** for all α's.

Remark : The Streda formula presented here is the mathematical version of a formula relating the derivative of the density of states with respect to the magnetic field to the Hall conductivity [see this relation in 18]. In the original work by Streda [89] another contribution was added to take into account the effect of the disorder in the crystal.

IV- Hofstadter like Spectra :

IV-1) The spectrum of the Harper operator-

In the section II we saw that the effective hamiltonian describing the quantum motion of a 2D Bloch electron in a uniform magnetic field can be expanded in a non commutative Fourier expansion in the operators U and V in the Universal Rotation Algebra. However in practice the exact expression may be quite complicate and if we only want to get qualitative properties, simplifications are required. The simplest approximation was proposed by Harper [49] in 1955 in dealing with square lattices, whereas Claro and Wannier [25] studied the corresponding model for triangular lattices (eventhough they claimed having studied the hexagonal one) and R.Rammal investigated recently the hexagonal lattice [81]. The square lattice is characterized by a complete symmetry of the hamiltonian under a $\pi/2$ rotation in space, which in turn means that it must be invariant under the automorphism β of $\mathcal{A}_\alpha$ defined by $\beta(U) = V$ and $\beta(V) = U^*$. If we retain in the Fourier expansion only the lowest order terms we end up with the Harper model :

$$H = U + U^* + V + V^* \tag{1}$$

Let us notice however that the physical meaning of this approximation is different at high and low magnetic field : in the former case it consists in approximating the potential itself by its first Fourier coefficients, whereas in the former case, it is the energy band function in momentum space which is approximated. Thus the strong field-weak field duality has also something to do with the position momentum duality by mean of Fourier transform.

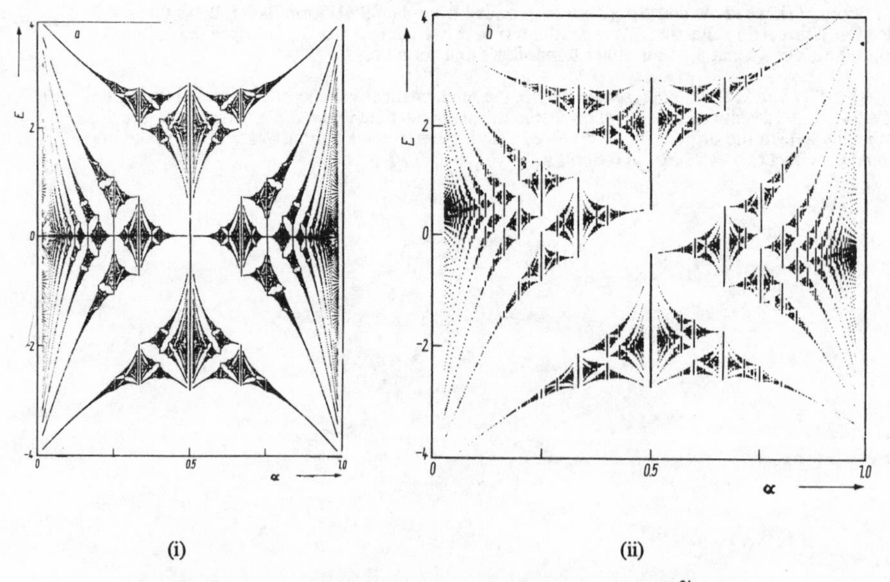

(i) (ii)

Fig.1- (i)-The spectrum of the Harper Hamiltonian $H = U+U^*+V+V^*$, *where* $UV = e^{2i\pi\alpha} VU$ *as a function of the parameter* α.
(ii)-The same for the Hamiltonian $H(s) = U + V + \sigma(e^{i\pi\alpha}V - e^{-i\pi\alpha}V^*)U$ + h.c. *with* $\sigma = 0.1$ *(Taken from Ref.* [26ii]*)*

This model has been the focus of attention ever since. Among the main contributions let us notice the work of M.Ya.Azbel' [12] in 1964 who guessed a rule for building up the spectrum. Independently in 1965, W.G.Chambers [24] using a semi classical approach proposed by Onsager [70], Adams II [2], Blount [21] and A.B.Pippard[78], rediscovered the rule together with the practical algorithm to compute numerically the spectrum. As in the work of Azbel' whose justification is still today quite obscure, he realized that if $\alpha = 1/(a_1+1/(a_2+...))$ is decomposed into continuous fraction expansion, each original band is divided into a_1 subbands, each of which being divided into a_2 subbands and so on. Nevertheless he "doubted whether this result is of great practical interest" an interesting comment in view of the future developments of the subject. D.Langbein [42, 64] gave a rather detailed calculation of the subband structure of (1) whithout getting far enough however to see the global structure in term of the magnetic field. D.G.Hofstadter [53] made it in 1976 and deserves the credit of having found a fascinating fractal structure (see fig.1i) by a time where it was no longer shameful to consider these monsters in Physics. Actually the precise rule in getting the subbands in the Hofstadter spectrum is a little bit different from the Azbel' and Chambers one, as was explained by Hofstadter : the middle bands have a different pattern but the rule may also be described in term of a modified continuous fraction expansion.

Let us also indicate that if one breaks the square isotropy, one gets the so called "Almost Mathieu" operator [14, 86]:

$$H(\mu) = U + U^* + \mu\,(V+V^*) \tag{2}$$

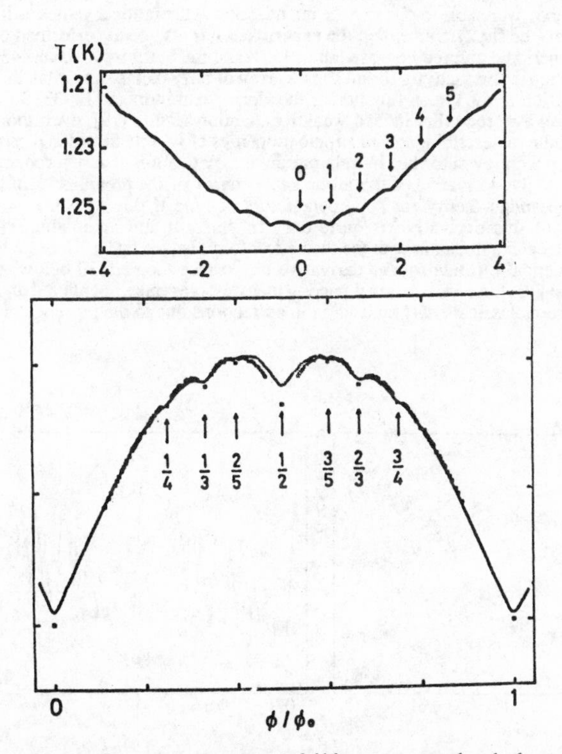

Fig.2- (upper fig.)-Critical temperature versus magnetic field for a square network made of superconducting aluminium. The arrows indicate the magnetic field values corresponding to integral number of flux quanta $\alpha = \phi/\phi_0$ per unit cell of the network. One sees a parabolic background corresponding to the bulk properties of the superconductor (see [100]*).*
(lower fig.)-A magnified view of the first period of the critical line in reduced units with a view of the secondary dips at α = 1/4, 1/3, 2/5, 1/2, 3/5, 2/3 and 3/4. The parabolic background has been removed. Dots represent the theoretical values calculated for rational p/q ($q \leq 30$). Taken from Ref. [73].

It has been originally introduced by S.Aubry [7, 8] in connection with the study of 1D organic conductors and their representation by mean of a quasi periodic potential. He then introduced the "Aubry duality" using an idea of Derrida and Sarma, which leads to the existence of a metal-insulator transition for the 1D representation of this hamiltonian. In algebraic terms Aubry's duality means that $H(\mu)$ and $\beta(H(\mu)) = \mu H(1/\mu)$ have the same spectrum (an elementary statement in our framework since β is a *automorphism). More precisely if E is the energy, the substitution $\mu \to 1/\mu$, $E \to E/\mu$ does not change the spectrum. It does not change the density of states either and through the Herbert-Jones-Thouless formula [52, 93] permits to show that for $\mu > 1$ the Lyapounov exponent is positive, leading (with some care [86] to localization [55, 74].

IV-2) The Grenoble experiments -

It turns out that the groundstate energy of the Harper hamiltonian and the corresponding models for other lattice symmetries, are describing the metal-superconductor transition curve for an array of superconductors in the magnetic field-temperature plane. The connection goes back to a work by P.G.de Gennes [40] and Alexander [3] in using Landau theory of superconductivity. In the case of a regular array, square or hexagonal lattice, Pannetier, Chaussy and Rammal [72] showed that the transition temperature for the superconductor in the array, submitted to a uniform magnetic field B, is given by a rather simple function of the groundstate energy $E(\alpha)$ of the Harper hamiltonian, where α is still defined as in §II eq.11.

They realized experimentally such an array with the hexagonal symmetry and were able to fit perfectly the numerically computed curve: in particular they found not only a flux quantization at integer values of α but they were able to observe the quantization at the rational values as it goes from the Hofstadter spectrum (see fig.2). Since then the experiment has also been performed on a square lattice [73], on a quasi periodic and the Penrose lattice [13], and the Serpinsky gasket [43]. Later on they realized that the electric conductivity for an square array of dirty metals was related through the theory of weak localization to the Green function of the Harper hamiltonian [32, 33, 34]. It gives a spectacular experimental evidence that indeed weak localization results [19], even though hardly rigorous, are quite accurate in describing the electronic properties of weakly disordered cristals.

More recently they went even farther by relating the magnetization of a superconductor in an array, at the transition, to the derivative of the groundstate energy of the previous hamiltonian [39, 99, 101] by using the Landau theory for superconductors of type II due to Abrikosov [1]. The numerical calculation of this derivative is quite easy to perform and again the experimental measurement is in spectacular agreement with the theoretical formula (see fig.3).

Actually the theoretical formula for the derivative $\partial E/\partial \alpha$ (see theorem 10 below) was firstly derived by M.Wilkinson [103] in a very original paper which deserves more attention than it got up to now. Using a delicate semiclassical analysis leading to corrections due to the presence of a "Berry's

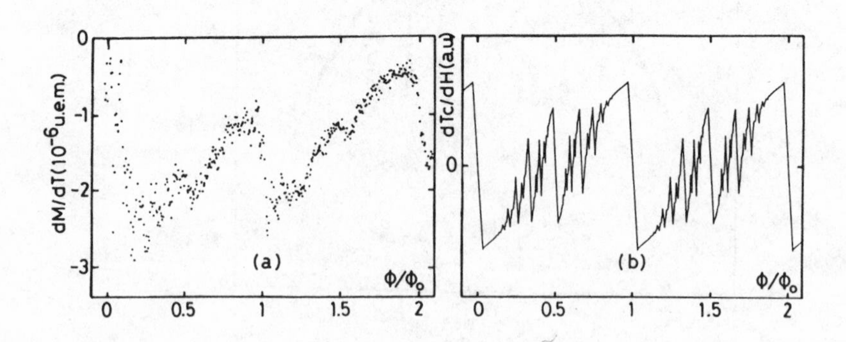

Fig.3- (a)-Experimental derivative of the magnetization dM/dT *versus the reduced flux* $\alpha = \phi/\phi_0$ *at* T = 3.05K *in a square network of superconducting indium.*
(b)-The numerical derivative dT_c/dH *of the theoretical mean field transition temperature.* dT /dH *is proportional to* $\partial E/\partial\alpha$. *Taken from Ref.* [39].

phase"[20, 87], he produced the exact value of the derivative of the gap boundaries with respect to the magnetic field. Rammal interpreted the result in term of the local density of state at the top of the band, near the rational values of the flux and remarked that it could be measured in term of the magnetization. For this reason we will call it the Wilkinson-Rammal formula.In the present section we are going to derive this formula for a C^k $(k>2)$ element of $\mathcal{A}$ by a completely different method than the one used by M.Wilkinson. Indeed a major defect of the Wilkinson calculation comes from the fact that it works only for polynomials of partial degree one in either U or V. It could probably be extended with some effort to polynomial of higher degree, but there is little hope that it could work for other examples. We will use instead our Ito derivation and a purely algebraic framework to derive it. However, prior to this calculation, we need a good knowledge of the weak magnetic field limit which can be performed via a semi classical analysis. To achieve this goal we have used ideas developped recently by P.Briet, J.M.Combes, P.Duclos [22, 28], and by B.Hellfer, J.Sjöstrand [50, 51].

IV-3) Semi-classical results-

Let $\mathcal{H}(\mathbf{k},\alpha)$ be a continuous function of the variables $\mathbf{k}=(k_1,k_2)\in\mathbf{R}^2$ and $\alpha\in(-\varepsilon,\varepsilon)$ for some $\varepsilon>0$. We assume that it satisfies the following properties:

(i) $\mathcal{H}$ is periodic with respect to $\mathbf{k}$ of period 2π in each component of $\mathbf{k}$.
(ii) If $\mathcal{H}=\Sigma_{\mathbf{m}\in\mathbf{Z}^2}\, h(\mathbf{m},\alpha)\, e^{i\mathbf{k}\wedge\mathbf{m}}$ is its Fourier expansion (where $\mathbf{k}\wedge\mathbf{m}=k_1m_2-k_2m_1$), then either :

$$\|\mathcal{H}\|_{(k)}=\sup_{i\le k}\Sigma_{\mathbf{m}\in\mathbf{Z}^2}\,|\partial^i h(\mathbf{m},\alpha)/\partial\alpha^i|(1+|\mathbf{m}|)^k<\infty \qquad \text{for some } k>2 \quad (3a)$$

or the $h(\mathbf{m},\alpha)$'s are holomorphic in α in a strip of width r and

$$|||\mathcal{H}|||_r=\sup_{|\mathrm{Im}(\alpha)|\le r}\Sigma_{\mathbf{m}\in\mathbf{Z}^2}\,|h(\mathbf{m},\alpha)|\,e^{r|\mathbf{m}|}<\infty \qquad \text{for some } r>0 \quad (3b)$$

(iii) For each α in $(-\varepsilon,\varepsilon)$ the function $\mathbf{k}\to\mathcal{H}(\mathbf{k},\alpha)$ has a unique regular minimum in each cell of period. Without loss of generality one can assume that this minimum is located at $\mathbf{k}=0$ for $\alpha=0$ and that $\mathcal{H}(\mathbf{0},0)=0$.

Correspondingly we define the quantized of $\mathcal{H}$ as the following element of $\mathcal{A}_\alpha$:

$$H(\alpha)=\Sigma_{\mathbf{m}\in\mathbf{Z}^2}\, h(\mathbf{m},\alpha)\, W(\mathbf{m}) \qquad W(\mathbf{m})=e^{i\pi\alpha m_1m_2}\, U^{m_1}\, V^{-m_2} \quad (4)$$

The groundstate energy $E(\alpha)$ is defined as the infimum of the spectrum of $H(\alpha)$ in $\mathcal{A}_\alpha$. Our first result concerns the asymptotic behaviour of $E(\alpha)$ as $\alpha\to 0$ namely :

Theorem 8 : Let $\mathcal{H}$ satisfy (i),(ii),(iii) and let $H(\alpha)$ be given by (4). Then the groundstate energy of $H(\alpha)$ is given by :

$$E(\alpha)=2\pi|\alpha|\sqrt{\det\{\frac{1}{2}\frac{\partial^2 H}{\partial k_i\partial k_j}(0,0)}+\alpha\frac{\partial H}{\partial\alpha}(0,0)+O(|\alpha|^{\delta/2}) \quad (5)$$

where $\delta=\mathrm{Min}(3,k)$.

◊

The proof of this result is actually an immediate consequence of the following result :

Theorem 9 : Let $\mathcal{H}$ satisfy (i),(ii),(iii) and let $H(\alpha)$ be given by (4). Then there is $E_c>0$ and $\varepsilon_c\le\varepsilon$ depending only on $\mathcal{H}$ such that if $\alpha\in(-\varepsilon_c,\varepsilon_c)$ the spectrum of $H(\alpha)$ below E_c is contained in the union of the intervals $\Sigma_n=[E_n(\alpha)-\delta(\alpha),E_n(\alpha)+\delta(\alpha)]$ where if $\mathcal{H}$ satisfies (3a) :

$$E_n(\alpha)=(2n+1)2\pi|\alpha|\det^{1/2}\{1/2\, D^2\mathcal{H}(\mathbf{0},0)\}+\alpha\,\partial\mathcal{H}/\partial\alpha(\mathbf{0},0)+O(|\alpha|^{\delta/2}) \quad (6)$$

$$0<\delta(\alpha)\le C_1|\alpha|^{k/2} \quad (7)$$

where C_1 is a constant depending only on $\mathcal{H}$.
If $\mathcal{H}$ satisfies (3b), the estimate (7) is replaced by :

$$0 < \delta(\alpha) \leq C_1 \, e^{-C_2/\alpha} \qquad (8)$$

where $C_2 \leq r$. ◊

The proof of this result is too long to be even sketched here eventhough there is no more ideas than in the semiclassical analysis by Briet-Combes-Duclos [22, 28] and Hellfer-Sjöstrand [50, 51]. Let us indicate however that the estimates (7) and (8) are in some sense optimal : from the proof it is possible to produce examples of $\mathcal{H}$'s for which they cannot be improved. However if $\mathcal{H}$ is a trigonometric polynomial, C_2 can probably be estimated by mean of a tunneling action (called also Agmon's distance). The difficulty here is that because of the symetric roles played by the components of **k**, we must perform a semiclassical analysis in phase space, using a canonical formalism instead of using a special choice of canonical coordinates. The question of how to get the best estimate for C_2 without further assumptions on $\mathcal{H}$ is still under progress. Let us note however that Hellfer-Sjöstrand [51] computed this constant in the case of Harper's model, using special positivity properties of its quantization.

Looking at Hofstadter spectrum (fig.1i), the Claro ones (fig.1ii) or at spectra computed by M.Wilkinson [104] for other models, we see easily the bundles of "quasi" eigenvalues given by the E_n's emerging from the bottom of the spectrum at small α's.

IV-4) The Wilkinson-Rammal formula-

One of the main property of the Hofstadter spectrum is its self-reproducing property at different scales. This suggests a Renormalization Group analysis. Such a method has been actually successfully used by several authors like Sokoloff or Economou and Soukoulis [see 88], in dealing with the metal insulator transition for the Almost Mathieu operator. It has been more systematically used for the computation of the spectrum by I.M. Suslov [90], S. Ostlund et al. [71], and by M.Wilkinson [104]. In their recent rigorous work on Harper's equation Hellfer and Sjöstrand[51] followed quite closely the Wilkinson strategy, and were able to prove rigorously that indeed each band edge can be decomposed into subbands according to the heuristic scheme, and that the restriction of the hamiltonian on each such subband can be analyzed in the same way. The main difficulty lies in the nature of the energy spectrum in the middle of each subband where the classical orbits are close to the separatrix and the semiclassical analysis must be modified (This is related to the "magnetic breakthrough" or "breakdown"[21, 78]) : for this reason Hellfer and Sjöstrand have been forced to eliminate the center of each subband from their analysis. They have also technical constraints on the continuous fraction expansion of α, which lead to a set of diophantine numbers of zero Lebesgue measure. For this reason one cannot conclude yet that the spectrum is a Cantor set except in special cases [15]; neither can we say whether its Lebesgue measure vanishes as predicted by the numerical work of S.Aubry and G.André [8], by the Thouless inequality and his numerical estimates[96], nor whether its Hausdorff dimension agrees with the recent predictions of Tang and Kohmoto [92] that it be equal to 1/2 .

We will be less ambitious here and will use a weaker version of the Renormalization Group method. The idea is that whenever α is close to a rational number p/q, the algebra $\mathcal{A}_\alpha$ can be identified with the subalgebra of $M_q \otimes \mathcal{A}_{\alpha-p/q}$ generated by the elements :

$$U_\alpha = W_1 \otimes U_{\alpha-p/q} \qquad\qquad V_\alpha = W_2 \otimes V_{\alpha-p/q} \qquad (9)$$

where W_1 and W_2 has been defined inTheorem 4 (§III eq.2) and the lower index in U and V indicates the rotation number of the corresponding algebra. Using the representation (9), one can write the hamiltonian as a matrix with entries in $\mathcal{A}_{\alpha-p/q}$. For α=p/q, the entries become function of the variable **k** in the 2-torus (theorem 4). Therefore, the spectrum can be computed simply by diagonalizing the corresponding matrix for each **k**, giving rise to band spectrum. Thanks to Elliott's theorem 5, a gap at α=p/q will survive for α in a small neighbourhood of p/q. Therefore the band hamiltonian will be computed by mean of a Cauchy formula on the resolvent :

$$H_B = \int_\gamma \frac{dz}{2i\pi} \, z \, \frac{1}{z1 - H} \qquad (10)$$

where γ is a contour surrounding once the band B we are looking at. We will assume here that B is not degenerate, namely that if P denote the eigenprojection of the band at the value $\alpha=p/q$, then $tr(P(\mathbf{k})) = 1$ for all $\mathbf{k}$'s. Let tr_q represents the partial trace induced by the usual trace on M_q on $M_q \otimes \mathcal{A}_{\alpha-p/q}$. We need the following lemma :

Lemma 1 : Let $E(\alpha)$ be the lower (resp. the upper) edge of the band B, and let $E'(\alpha)$ be the lower (resp. the upper) edge of the spectrum of the element $tr_q(H_B)$ of $\mathcal{A}_{\alpha-p/q}$. Then :

$$E(\alpha) = E'(\alpha) + O((\alpha - p/q)^2) \tag{11}$$

◊

The proof of this lemma uses again the projection method as in the proof of the validity of the Peierls substitution (§II-2).
We now remark that the element $tr_q(H_B)$ of $\mathcal{A}_{\alpha-p/q}$ coincides at $\alpha=p/q$ with the energy band function $E_B(\mathbf{k})$ and the Peierls substitution consists in identifying $E(\alpha)$ with $E'(\alpha)$. If we apply now the theorem 8 to $tr_q(H_B)$ we need first to compute the matrix of second derivatives of $E_B(\mathbf{k})$ at the bottom (resp. the top) of the band. However, as remarked by R.Rammal [see 101], the determinant $\det^{1/2}\{1/2\ D^2E_B(\mathbf{0})\}$ is nothing but q^2 time the inverse of the local density of state $\rho(E_B(\mathbf{0}))$at the corresponding band edge. Secondly, we must compute the value at $\alpha=p/q$ of $\partial tr_q(H_B)$ evaluated at the corresponding band edge. We have done it by using the theorem 7 in calculating the Ito derivative of a resolvent. The result is the following:

Theorem 10 : Let H satisfy (i),(ii),(iii) above with $\partial H=0$, $k \geq 3$, and let B be a non degenerate band of H at $\alpha=p/q$. The lower (resp. the upper) edge of the band $E^-(\alpha)$ (resp. $E^+(\alpha)$) is given by the Wilkinson Rammal formula :

$$E^{\pm}(\alpha) = E^{\pm}(\tfrac{p}{q}) - (\pm)\, a\, |\alpha - \tfrac{p}{q}| + b\, (\alpha - \tfrac{p}{q}) + O(|\alpha - \tfrac{p}{q}|^{3/2}) \tag{12}$$

with :

$$a = \frac{2\pi q^2}{\rho(E^{\pm}(\frac{p}{q}))} \quad ; \quad b = \frac{E^{\pm}(\frac{p}{q})}{4i\pi}\, tr_q\{P(\mathbf{k})[\partial_1 P(\mathbf{k}), \partial_2 P(\mathbf{k})]\}\Big|_{E_B(\mathbf{k})=E^{\pm}(\frac{p}{q})} \tag{13}$$

◊

The first term in this formula is therefore the value of the energy at the band edge for $\alpha=p/q$. The second represents the harmonic oscillator effect, and it produces a discontinuity in the derivative. It shrinks the spectrum in such a way that the neighboring gap actually increases in size due to this term. The last term comes from Berry's phase, namely from the fact that the eigenprojection $P(\mathbf{k})$ at the value $\alpha=p/q$ defines in general a non trivial line bundle over the 2-torus [20, 87]. We notice however that this second term is *not* the Chern class of the bundle for the Chern class is obtained firstly in dividing the trace by q and then integrating over the 2-torus (Brillouin zone). Here we do not integrate, but we rather evaluate the integrand at the band edge. This last term accounts for a dissymetry of the derivative around $\alpha=p/q$ and may partially distroy the effect of the first one on the enlargement of the corresponding gap. The derivative of the magnetization of the superconducting array at the transition with respect to the temperature is actually a simple function of the dissymetry of the derivative at each rational point (see [99]).

The previous theorem is established for an element H such that $\partial H=0$. If $\partial H \neq 0$ there is an additional contribution to the second term which we will not give here but which is easy to compute.

One consequence of this formula is the following :

Corollary : Let $E(\alpha)$ be a gap boundary for $H \in \mathcal{A}_\alpha$. For any irrational value of α, $E(\alpha)$ is differentiable.

◊

Open problem: Is it possible from this formula to get a proof that the spectrum of H is actually a Cantor set for any irrational α ?

V- The Integer Quantum Hall Effect:

V-1) The physical situation-

In 1880, Hall [46] performed an experiment which led to the discovery of what has been called the Hall effect. A century later in 1980, Von Klitzing and his collaborators [57] showed that at very low temperature the Hall conductivity of a two dimensional disordered crystal exhibits a quantization at multiple values of the fundamental constant e^2/h with a very high accuracy [See also 97, 98]. After the works of Laughlin [65] and especially of Thouless, Kohmoto, Nightingale, Den Nijs [95], it became clear that indeed the quantization of the Hall conductivity at low temperature had a geometric origin. However the conventional framework used by Solid State physicists provides strong limitations : in their work Thouless, Kohmoto, Nightingale, Den Nijs were forced to consider perfect crystals with rational magnetic fluxes. Neither could they take into account the disorder which is of extreme importance in creating the effect [5, 48, 79, 84, 94], nor could they avoid the rationality of the magnetic flux, a rather unphysical hypothesis. The C*Algebra approach that we already described helps actually to solve these two difficulties. It turns out that the Non Commutative Geometry developped by A.Connes [29] constitutes a natural framework to give a rigorous description of the Quantum Hall Effect and to prove that in the framework of ordinary Quantum Mechanics, the Hall conductivity is indeed quantized in the zero temperature limit. The purpose of this last section is to complete the set of arguments presented in [18]. However we want also to emphasize that a very similar argument not using that sophisticated language has been given by H.Kunz [62] and gives rise to results in much better agreement with the physical situation than the previous ones eventhough for technical reasons, H.Kunz is forced to assume the existence of gaps in the energy spectrum which are not observed in the devices used in practice. We will offer a solution to this problem also.

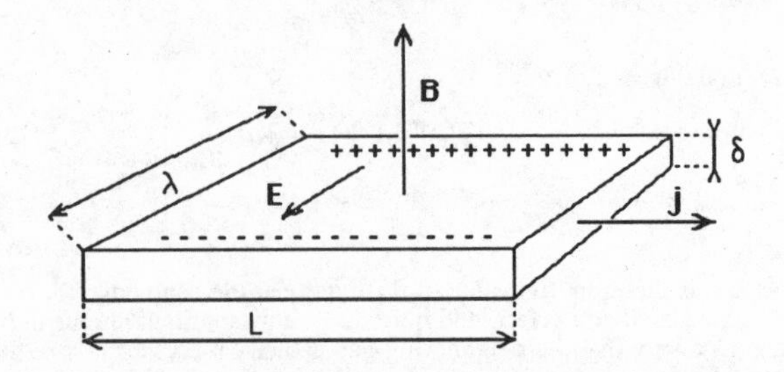

Fig. 4 : The classical Hall effect. The sample must be a thin strip of pure metal. The magnetic field B is perpendicular to the strip. The electric current j is stationnary and oriented along the strip in the direction of the longest side. The magnetic forces push the charges as indicated. In the stationnary state, they create an electric field E perpendicular to the directions of the magnetic field and the current. One measures the Hall voltage in the E-direction.

The Hall effect concerns a two dimensional gas of classical charged particles submitted to a uniform magnetic field **B** (cf.fig.-4). When a current density **j** is created in the bulk of the system, it reacts in creating an electric field **E** such that the force acting on it vanishes in the stationnary state giving rise to the relation :

$$N e \mathbf{E} + \mathbf{j} \wedge \mathbf{B} = 0 \qquad (1)$$

where N is the charge carriers density and e the charge of the carriers. Thus **E** is perpendicular to **j** and **B** and **j** is linear with respect to **E**. Solving (1) leads to the definition of the Hall conductivity σ_H :

$$\mathbf{j} = N e \mathbf{B} \wedge \mathbf{E}/B^2 \quad \Rightarrow \quad |\mathbf{j}| = j = \sigma_H |\mathbf{E}| \quad \text{with} \quad \sigma_H = Ne/B \qquad (2)$$

We see in particular that the Hall voltage created in the direction perpendicular to the magnetic field and to current depends upon the sign of the charge carriers. As soon as 1880, Hall observed that this sign may be negative or positive depending upon the metal used for the experiment. This was the first experimental evidence of what is nowadays called hole or electron conduction. It is still today a currently used method. The Hall conductivity varies also with the charge carrier density N, a property which is used in modern devices as the MOSFET or AsGa heterojunctions [4]: in practice one can keep B fixed and quite large ≈18 T, while one varies N by tuning the gate voltage [57], whereas in other experiments the magnetic field has been used as physical parameter [97].

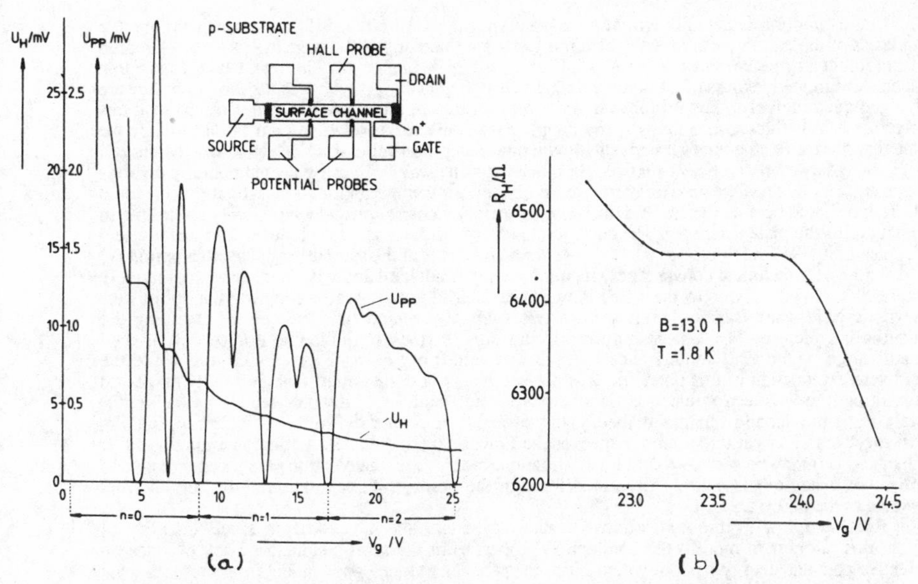

Fig. 5 : (a) Recording of the Hall voltage U_H *and the voltage drop between the potential probes* U_{pp} *as a function of the gate voltage at* T=1.5 K*. The magnetic field is* 18T*. The oscillation up to the Landau level n=2 is shown. The Hall voltage and* U_{pp} *are proportional to the conductance* ρ_{xy} *and* ρ_{xx} *respectively. The inset shows a top view of the device with a length of* L=400μm *a width of* W=50μm *and a distance between the potential probes of* L_{pp}=130μm.
(b) A detailed view of a plateau . (Taken from Ref.[57]*)*

In their experiment Von Klitzing, Pepper and Dorda worked at very low temperature and they observed that the Hall conductivity is not a linear function of the charge carriers density, but exhibits some plateaux (see fig.5) where it takes values equals to a multiple integer of the physical constant e^2/h with a very high accuracy (less than 10^{-7} in the most recent experiments). At the same time they observed that values of the gate voltage giving rise to these plateaux correspond to vanishing of the direct conductivity, indicating that the Fermi level in the sample is lying in a region of localized states. An explanation of this phenomena was first given by Laughlin [65] in 1981, using the gauge invariance of the one electron hamiltonian. Later on Avron and Seiler improved it [10], by giving a mathematically rigorous argument that the Hall conductivity should be quantized whenever the Fermi level lies in an energy gap. In these approaches however the topology of the sample in space seems to play a crucial role and they do not explain the role of the localization in producing plateaux.

Still, many physicists are nowadays unsatisfied with these arguments. Experiments using microwave fields exhibit also a quantization of the Hall conductivity without having the topology required by that arguments[61]. Moreover, while localized states should contribute to the charge carrier density, they should not contribute to the current [79, 94]. A microscopic theory insensitive to the shape of the sample by considering the infinite volume limit should be closer to the real situation. We want to emphasize here that according to the argument of Thouless, Kohmoto, Nightingale & Den Nijs [95] *the origin of the topological effect is not the shape of the sample, but rather the topology of the Brillouin zone, in momentum space.* But the word "topology" will be understood here in the non commutative sense instead. Our argument shows that localized states give no contribution to the Hall conductivity, giving rise to the existence of plateaux at exactly the values observed in the experiments. It is also independent upon the detail of the one particle hamiltonian and the value of the magnetic field. In particular it also works in the tight binding representation on every kind of lattices [see 16, 18, for some informations].

The assumptions required for this result to hold are the following :
(i) the sample is two dimensional.
(ii) the sample is considered at the infinite surface limit.
(iii) the temperature is zero.
(iv) the current flowing into the sample is so small as to be neglected.
(v) one neglects the spin of the electrons.
(vi) one neglects the electron-electron interaction.

The first asumption is actually essential in observing the Hall effect [46]. In heterojunctions the thickness of the region containing the electron gas is approximatly 200Å large in samples of hundred of micrometers in size. On the other hand, by changing our algebra, one could take easily into account the finite thickness of the sample without changing qualitatively the results (however they are changed quantitatively). The infinite surface limit may appear more questionable. But observations performed show fluctuations may distroy the plateaux if the size is relatively too small [56]. On the other hand these fluctuations get very small with increasing the sample size, in such a way that indeed (ii) is necessary, and practically realized. In much the same way, increasing the temperature distroys the quantization. Thus the zero temperature limit is the convenient region to investigate. With some effort one should be able to compute the finite temperature corrections which probably contribute to rounding off the plateaux edges. The current must be small. Increasing it produces non linear effect like bistability [91] but not too small however for a given size of the sample [56]. The assumption (v) is there only for technical convenience. Including spins is not hard and will not change qualitatively the results. However, due to the Pauli term in the Schrödinger operator coupling the spin to the magnetic field, each Landau bands split in two subbands at high field (see fig.5), changing the quantitative picture. The last assumption is the most serious limitation to our theory. Indeed quantization of the Hall conductance at fractional values has been observed [98] and while the complete explanation of this phenomena is not yet understood, the scientific community agrees on attributing it to the electron-electron interactions. In our framework, it would require extending the analysis to the second quantized theory and probably need the definition of a "multiplicative K-theory" which is yet to be done. However the Fractional Quantum Hall effect is a much weaker effect and (vi) may be seen as a first approximation. In this case, the electrons are independent, and collective behaviour is entirely described through the Fermi statistics in computing the thermal average of the observables.

Real samples, eventhough almost perfect crystals, have necessarily a small quantity of impurities in order to modify the conduction properties of the semi conductor. They produce an effective potential energy randomly distributed in space. Therefore a good model for representing the electron hamiltonian is given by :

$$H = H_L + V_\omega(X) \tag{3}$$

where the second term represents the potential created by the disorder and ω the configuration of the disorder : ω will be a random variable. Because that potential energy is much bigger in size than the distance between two Landau levels, for most sample used in practice the spectrum of H has no gap. The density of states ρ(E) is defined as the derivative with respect to the energy of the number of states below the energy E per unit volume. In real samples it looks like in fig 6, namely it is positive and the ratio between the minimum and the maximum may be as big as 30% (see Ref. [45] for more details and references). In this case the parameter N/B becomes a monotone increasing function of the chemical potential μ which coincides at low temperature with the Fermi energy E_F. For this reason mathematician is allowed to consider the Fermi energy E_F as the variable parameter instead of N/B. Thus the absence of gaps in the energy spectrum due to the presence of a strong impurity potential is necessary to replace the physical parameter N/B by a parameter involving only spectral properties of the one particle hamiltonian. The reader is invited to look at [18] for a more complete discussion of the localization problem.

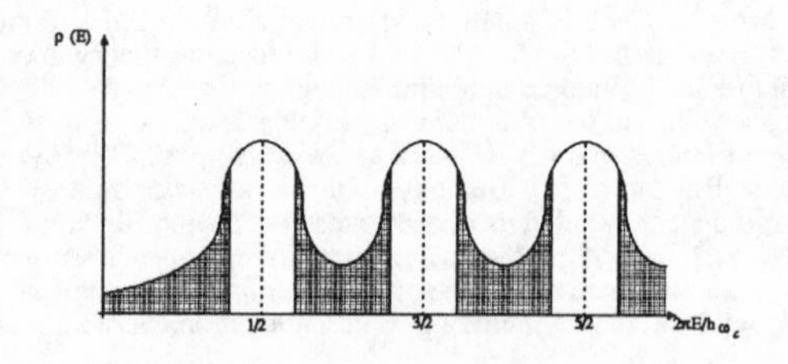

Fig.6- The density of states for a typical sample in a uniform magnetic field. The energy is given in unit of the Landau energy. The dashed area corresponds to localized states. In some samples the region of extended states can be invisible. The minimum of the density of states can be as big as 30% of the maximum (see Ref.[45] for details).

V-2) The Observable Algebra-

In the previous section, we introduced a description of the disorder through the configuration ω and the impurity potential. The main property of an infinite surface limit is that the sample looks homogeneous, namely it has the same properties if one changes the origin of the coordinates. More precisely, let Ω be the set of impurity configurations. Since the disorder is random, it is reasonnable to assume that Ω is a probability space. On the other hand, moving the sample is equivalent to moving the configuration. Therefore there is an action $\{T^{\mathbf{a}}: \mathbf{a} \text{ in } \mathbf{R}^2\}$ of the translation group in Ω. One will assume that this probability is invariant (homogeneity of the sample) and ergodic under this action. Let us consider the hamiltonian H as a measurable function (in the strong resolvent sense) of the impurity configuration . The homogeneity gives :

$$U(\mathbf{a})\, H_\omega\, U(\mathbf{a})^* = H_{T^{\mathbf{a}}\omega} \qquad \mathbf{a} \in \mathbf{R}^2 \tag{4}$$

where U(a) represents the action of the translation by the vector **a** of $\mathbf{R}^2$ in the Hilbert space $L^2(\mathbf{R}^2)$ of wave functions. Actually for the model described in the eq.1, (4) is satisfied only if U(a) is the so-called magnetic translation operator [105] defined by:

$$U(\mathbf{a})\, \psi(\mathbf{x}) = e^{i\pi B\mathbf{x}\wedge\mathbf{a}/\phi}\, \psi(\mathbf{x}-\mathbf{a}) \quad \phi = h/e \qquad \psi \in L^2(\mathbf{R}^2) \tag{5}$$

where $\mathbf{x}\wedge\mathbf{a} = x_1a_2 - x_2a_1$ if $\mathbf{x} = (x_1,x_2)$ and $\mathbf{a} = (a_1,a_2)$. Once and for all we will set $\lambda = \pi B/\phi$.

Physically any of the translated of the hamiltonian gives an equivalent description of the system. For this reason, the observable algebra must contain the whole family $\underline{R}(z) = \{(z-H_\omega)^{-1};\ \omega \in \Omega\}$ for some non real z. As we see *this family is not commutative* ! We will denote by $\mathcal{A}(B)$ the C*Algebra generated by the family $\underline{R}(z)$. It turns out that one can compute this algebra for it is a special case of the algebras studied by J.Renault [109].

To this aim we introduce on Ω a topology of a Hausdorff compact space in a natural way [16]. Considering the expression (3) the disorder appears through the potential energy which we will suppose bounded for simplicity (Probably relative boundedness with respect to H_L may be sufficient). Given V(X) in $L^\infty(\mathbf{R}^2)$, let us consider the weak closure Ω' in $L^2(\mathbf{R}^2)$ of the set $\{V(X-a);\ a \text{ in } \mathbf{R}^2\}$. Ω' is called the *hull* of V. It is a compact space, on which the translation group acts in an obvious way through the projective representation U. Since, from the point of view of Quantum Mechanics, the data of a disorder is equivalent to the data of the potential energy operator V(X), one can identify the set Ω of configurations of impurities to Ω'. The randomness will simply be described through a translation invariant ergodic probability measure **P** on Ω'. The virtue of this representation is contained in the following result:

Theorem 2 : If V is a bounded operator on $L^2(\mathbf{R}^2)$ and B the magnetic field, we let $H_{L,V}$ be the self adjoint operator (eq.II-1) :

$$H_{L,V} = H_L + V \tag{6}$$

Then the matrix element $< \mathbf{0} \mid (z-H_{L,V})^{-2} \mid \mathbf{x} >$ of the square of the Green function is continuous with respect to (B,V,x) if one endows the set of bounded operators V with the weak topology.

◊

This result together with the covariance property (4) allows us to built the algebra $\mathcal{A}$ as follows : let Ω be a now compact space with an action T of $\mathbf{R}^2$ through a group of homeomorphisms, and let **I** be a translation invariant ergodic measure on Ω. Let $\mathcal{A}_0$ be the set of complex valuedfunctions A of $(\lambda,\omega,x) \in \mathbf{R}\mathrm{x}\Omega\mathrm{x}\mathbf{R}^2$ continuous with compact support. We then define on it a product and an adjoint compatible with the formula (4) as follows :

$$AB(\lambda,\omega,x) = \int d^2y\ A(\lambda,\omega,y)\ B(\lambda,T^{-y}\omega,x-y)\ e^{i\lambda x\wedge y} \tag{7}$$

$$A^*(\lambda,\omega,x) = A\,(\lambda,T^{-x}\omega,-x)^*$$

This defines an algebra $\mathcal{A}_0(B)$ for any $\lambda = \pi eB/h$. One represents it as an algebra of operators on $L^2(\mathbf{R}^2)$ through the family $\underline{\pi}_\lambda = \{\pi_{\lambda,\omega};\ \omega \in \Omega\}$ of *-representations as (see eq. 4) :

$$\pi_{\lambda,\omega}(A)\ \psi(x) = \int d^2y\ A(\lambda,T^{-x}\omega,y-x)\ e^{i\lambda x\wedge y}\ \psi(y) \qquad \psi \in L^2(\mathbf{R}^2) \tag{8}$$

One can now define two C*-norms through this family $\underline{\pi}_\lambda$ by :

$$\| A \|_\lambda = \sup_{\omega\in\Omega} \| \pi_{\lambda,\omega}(A) \|$$

$$\|A\| = \sup_{\lambda\in\mathbf{R}}\|A\|_\lambda \tag{9}$$

Then $\mathcal{A}(B)$ (resp. $\mathcal{A}$) is the completion of $\mathcal{A}_0(B)$ (resp. of $\mathcal{A}_0$) under the norm $\|\,.\,\|_\lambda$ (resp. $\|\,.\,\|$). One consequence of this description and of the theorem 2 is:

Corollary : If V is a bounded operator on $L^2(\mathbf{R}^2)$ we let Ω be the hull of V. If $\omega\in\Omega$ we let V_ω be the multiplication operator determined by ω on $L^2(\mathbf{R}^2)$. The resolvent of the operator $H_\omega=H_L+V_\omega$ defines an element of the algebra $\mathcal{A}$.

◊

An integration theory is defined on $\mathcal{A}$ or $\mathcal{A}(B)$ through a trace τ as follows:

$$\tau(A)(\lambda) = \int \mathbf{P}(d\omega)\, A(\lambda,\omega,\mathbf{0}) \quad \Rightarrow \quad \tau(AB)=\tau(BA) \qquad A,B \in \mathcal{A} \tag{10}$$

It turns out that using Birkhoff's theorem [47, see 16], this trace is nothing but the trace per unit area, namely

$$\tau(A)(\lambda) = \lim_{V\to\infty} V^{-1}\,\mathrm{Tr}_V(\pi_{\lambda,\omega}(A)) \qquad A\in\mathcal{A} \tag{11}$$

At last a differential structure is defined through the pair ∂_i (i=1,2) as follows :

$$\partial_i A(\lambda,\omega,x) = 2i\pi\, x_i\, A(\lambda,\omega,x) \tag{12}$$

One checks immediatly, using (7) that it is indeed a *derivation . Moreover it satisfies :

$$\pi_{\lambda,\omega}(\partial_i A) = 2i\pi\,[X_i, \pi_{\lambda,\omega}(A)] \tag{13}$$

where $\mathbf{X} = (X_1,X_2)$ is the position operator. In particular, using a Fourier transform, one sees that it is a *derivative in momentum space only,* and therefore the geometry described here lies in momentum space and not in the the real space! In other words, $\mathcal{A}$ describes the *non commutative Brillouin zone* for our system!

We will denote by $C^k(\mathcal{A})$ (for k in N or k=∞) the set of elements of $\mathcal{A}$ in the domain of any k^{th} power of $\partial = (\partial_1,\partial_2)$. It can be shown that it is a dense subalgebra stable by smooth functional calculus [see 29 and 107]. As a generalization of theorems 2 and 5 above, we get [107]:

Theorem 12 : If A is an element of the C*Algebra $\mathcal{A}$, then $\|A\|_\lambda$ and $\tau(A)(\lambda)$ are continuous functions of λ. If A is self adjoint, the gap boundaries of its spectrum are also continuous functions of λ.

◊

V-3) On Kubo, Chern and Connes: practical formulæ -

The quantum calculation of the Hall conductivity starts with the calculation of the thermal average of the current. However there are quite a lot of difficulties because one has to assume some mechanism for the dissipation of the energy by mean of collisions. This part is a hard mathematical question and yet there is no definite formalism to justify that description from the microscopic dynamic. However physicists have satisfactory phenomenological theories giving rise to reasonable formulæ for the transport coefficients [68]. The so-called Kubo formula gives in particular an expression for the electric conductivity in term of microscopic data [59, 60].

In the case of the Hall conductivity however, the derivation we gave in §IV-1 for a free electron gas, shows that no dissipative mechanism is required to get a linear relation between the current and the Hall field. For this reason the proof of the Kubo formula in that case must be much easier for one can neglect collision in the first approximation. Indeed Avron, Seiler and Yaffe [11] gave a proof of it at zero temperature whenever one assumes that the Fermi energy lies in a gap of the energy spectrum. Actually, this assumption looks more like a technical convenience than an essential hypothesis: there is some hope that this proof can be extended to the case where the Fermi level lies in a region of localized states. In any case Kubo's formula becomes very simple at zero temperature in our algebraic framework, namely (omitting the explicit dependence in the magnetic field):

$$\sigma_H = \frac{e^2}{h}\frac{1}{2i\pi}\,\tau(P_F\,[\partial_1 P_F,\partial_2 P_F]) = \frac{e^2}{h}\,\mathrm{Ch}(P_F) \tag{14}$$

where P_F is the eigenprojection of the hamiltonian on energies smaller than or equal to the Fermi level. Thus we get the Japanese-Chinese relation **Kubo = Chern** at zero temperature!

As pointed out in §III-3 the right hand side of (14) is defined only for those projection in the Sobolev space $\mathcal{H}_1(\mathcal{A})$ of the algebra. Therefore we are faced with the problem of interpreting the condition that P_F be in $\mathcal{H}_1(\mathcal{A})$. Since it is the eigenprojection of an operator with resolvent in $\mathcal{A}$, it is trace class (with respect to τ) for it projects on a bounded interval of energy since the hamiltonian H is bounded below. In particular it is Hibert Schmidt. It remains therefore to check that the derivatives are also Hibert-Schmidt. Using eq.10 we get in fact :

$$\tau((\partial_1 P_F)^2 + (\partial_2 P_F)^2) = \int_\Omega \mathbf{dP}(\omega) \int_{\mathbf{R}^2} \mathbf{d}^2\mathbf{x}\, |<0| P_{F,\omega} |\mathbf{x}>|^2 |\mathbf{x}|^2 = \xi \qquad (15)$$

In some sense ξ measures the localization length at the Fermi level. Therefore *the finiteness of ξ means that the Fermi level belongs to the localized states region* and at the same time it gives a sufficient condition for the Chern class in (14) to be well defined.

To get the Hall conductivity quantized it is therefore enough to show that the Chern class of P_F is an integer. As it has been derived in [18] this is possible through using a formula due to A. Connes according to which the Chern class of a projection is given by the Index of a Fredholm operator. Let us quickly describe the construction.

As for computing the 2nd Chern class of a line bundle over a manifold one is led to construct a Dirac operator. This is done by doubling the number of components of the bundle. In much the same way, we double the Hilbert space to get $\mathbf{H} = \mathbf{H}_+ \oplus \mathbf{H}_-$ where $\mathbf{H}_\pm = L^2(\mathbf{R}^2)$. We define on $\mathbf{H}$ the following operators :

$$G = \begin{vmatrix} 1 & \\ & -1 \end{vmatrix} \qquad F = \begin{vmatrix} & X/|X| \\ X^*/|X| & \end{vmatrix} \qquad A = \begin{vmatrix} \pi_{\lambda,\omega}(A) & \\ & \pi_{\lambda,\omega}(A) \end{vmatrix} \qquad A \text{ in } \mathcal{A} \qquad (16)$$

where $X = X_1 + i X_2$ is the complexified position operator. Note that F is the Fourier transform of the phase of the Dirac operator in two dimensions, and that $\mathbf{H}$ will be the non commutative analog of the space of square integrable sections of a two–dimensional hermitian bundle over the "pseudo manifold" represented by the observable algebra. The triple $(\mathbf{H},G,F)$ is called a Fredholm module [6] for it has the following properties (the last one requires a calculation, but can be proved as in [29]) :

(i) G anticommutes with F and commutes with the elements of the observable algebra. An element of degree 0 will be an operator on $\mathbf{H}$ commuting with G, whereas an element of degree 1 is an operator on $\mathbf{H}$ anticommuting with G.
(ii) $F^2 = 1$
(iii) if A belongs to $\mathcal{A}$, $[F,A]$ is in $L^p(\mathbf{H})$, for $p > 2$. Here, $L^p(\mathbf{H})$ is the Schatten ideal of bounded operators C on $\mathbf{H}$ such that $|C|^p$ is trace class [85].

With this construction we can prove the following theorem [18, 107]:

Theorem 13 : The Chern character of any projection P in $\mathcal{H}_1(\mathcal{A})$ is an integer. It is given by the *Connes formula* :

$$Ch(P) = Ind(F_P) \qquad \mathbf{P}\text{ - almost surely} \qquad (17)$$

where F_P is the restriction of $\{PFP\}_{+,-}$ to the space PH. ◊

To give a hint in the proof of this theorem let us introduce some technical tools. A graded commutator is defined as follows on the set of bounded operators on $\mathbf{H}$:

$$[T,T'] = TT' - (-)^{\deg(T)\deg(T')}\, T'T \qquad (18)$$

and a graded trace Tr_S is given by :

$$Tr_S\,(T) = 1/2 Tr\,(\,GF[F,T]) \qquad (19)$$

One also defines a character by the following trilinear map on $\mathcal{A}(B)$:

$$\tau\,(A_0, A_1, A_2) = 1/2i\pi\, \tau\,(A_0\,\{\partial_1 A_1\, \partial_2 A_2 - \partial_2 A_1\, \partial_1 A_2\,\}) \qquad (20)$$

in particular [29] :

(i) it is cyclic : $\tau\,(A_0, A_1, A_2) = \tau\,(A_1, A_2, A_0)$
(ii) it is closed : $\tau\,(A_0A_1, A_2, A_3) - \tau\,(A_0, A_1A_2, A_3) + \tau\,(A_0, A_1, A_2A_3) - \tau\,(A_3A_0, A_1, A_2) = 0$

Moreover, as in [29] we get the following *Connes formula* (again the magnetic field is implicit here) :

$$\tau (A_0, A_1, A_2) = \int_\Omega dP(\omega)\ Tr_S\ (A_0[F,A_1][F,A_2]) \tag{21}$$

One remarks that since [F,A] belongs to $L^p(H)$ for $p > 2$, the right hand side of (5) is well defined. If now P is a projection in $C^1(\mathcal{A}(B))$ one sees that :

$$Ch(P) = \tau\ (P,P,P) \tag{22}$$

An important remark lies in the following lemma [107] :

Lemma 2 : The formula (22) extends to every projection in the Sobolev space $\mathcal{H}_1(\mathcal{A})$.

◊

The main ingredient in the proof of the integrality of the Chern character is the remark proved in [29, app.1,p.86] that :

$$Tr_S\ (P[F,P][F,P]) = Ind(F_P) = Ker\ (F_P) - Ker\ (F_P{}^*) = n(\lambda,\omega) \tag{23}$$

where F_P is the restriction of $\{PFP\}_{+-}$ to the space PH. Since F commutes to P up to a compact operator, and since $F^2 = 1$ it follows that F_P is invertible in PH up to compact operators, and therefore it is a Fredholm operator. In particular $Ind(F_P)$ *is an integer*! In fact this integer may depend upon the disorder. However thanks to the homogeneity condition (eq.4) one has (including the explicit dependence with respect to ω) :

$$Ind(\{P_\omega FP_\omega\}_{+-}) = Ind(\{P_{T^a\omega} U(a)FU(a)^*P_{T^a\omega}\}_{+-}) \tag{24}$$

$$P_{T^a\omega} U(a)FU(a)^*\ P_{T^a\omega} = P_{T^a\omega} F P_{T^a\omega} + K \tag{25}$$

where K is some compact operator. This is because the gradient of X/|X| converges to zero at infinity. It implies :

$$n\ (\lambda,\omega) = n\ (\lambda,T^a\omega) \in \mathbf{Z} \qquad \text{for all a in } \mathbf{R}^2 \tag{26}$$

Since the probability measure **dP** on Ω is ergodic invariant, it follows that n(λ,ω) is almost surely a constant and therefore we have proved that the Hall conductivity is an integer whenever ξ is finite.

It is even possible to compute this integer. Let H(s) be the operator obtained from the initial model (eq.3) by replacing the potential V by sV ($0 \leq s \leq 1$). H(s) is a norm resolvent continuous family of operators if $V \in L^\infty(\mathbf{R}^2)$. One can show from this property that $Ch(P_F(s))$ is also continuous in s [107] as long as ξ (eq.15) stays finite. Since it is integer valued it must be constant. In [18] we discussed the existence of localized state region. There are good evidences that in between two Landau levels there is always a small region of localized states. Therefore by varying E_F with s as well in order to keep it in a region of localized states while s varies from 1 to 0, the number n(λ,ω) is equal to its value in absence of disorder! If the free hamiltonian is given by the Landau model (§II eq.1) it is equal to n whenever the Fermi energy lies between $(2n+1)E_0$ and $(2n+3)E_0$, if E_0 is the groundstate energy of the Landau hamiltonian.

Theorem 14 : If the Fermi level E_F belongs to a gap of extended states [E,E'] in the spectrum of the operator given by (3) in the sense that ξ (eq.15) be bounded on it, the Chern Class of the Fermi projection P_F is an integer given by the Connes formula (17). Moreover if one assumes that whatever s in [0,1] H(s) has always such a gap in between two Landau levels, this integer is independent of s and given by the theory without disorder. This integer is therefore the same in each gap of extended states.

◊

Corollary (Kunz[62]) : Between two plateaux of the Hall conductivity, the localization length ξ must diverge.

◊

This last result expresses precisely the claim by Halperin [48] according to which there must be at least one extended state between two plateaux. Eventhough at the infinite surface limit the hamiltonian may have only point spectrum [17], excluding the existence of such state the localization length diverges producing states localized in a region bigger than the size of the sample. Such states will behave in practice like extended ones [84] producing a non zero direct conductivity (fig.5). It is moreover reasonable to believe that the localization length actually diverges exactly at the Landau energies [5] when the spectrum is pure point provided the probability **P** be invariant under the flip V→(−V) of the potential [106]. This has already been shown is special examples [62] and a general analysis should not be out of reach.

REFERENCES

[1] A.A. ABRIKOSOV, On the Magnetic Properties of Superconductors of the Second Group, *Sov. Phys. JETP*, 5, (1957), 1174-82.

[2] E.N. ADAMS II, Motion of an Electron in a Perturbed Periodic Potential, *Phys. Rev.*, 85, (1952), 41-50.

[3] S. ALEXANDER, Superconductivity of networks. A percolation approach to the effect of disorder, *Phys. Rev.*, B27, (1983), 1541-57.

[4] T. ANDO, A.B. FOWLER, F. STERN, Electronic properties of two dimensional systems, *Rev. Mod. Phys.*, 54, (1982), 437-672.

[5] H. AOKI, T. ANDO, Critical Localization in Two-Dimensional Landau Quatization, *Phys. Rev. Letters*, 54, (1985), 831-834 and references therein.

[6] M. ATIYAH, K–Theory, *Benjamin, New-York, Amsterdam*, (1967).

[7] S. AUBRY, The new concept of transition bybreaking of analyticity in cristallographic models, *Solid State Sci.*, 8, (1978), 264.

[8] S. AUBRY, G. ANDRE, Analyticity breaking and the Anderson localization in incommensurate lattices, *Ann. Israel, Phys. Soc.*, 3, (1980), 133.

[9] J. AVRON, B. SIMON, Stability of gaps for periodic potentials under variation of a magnetic field, *J. Phys.*, A18, (1984), 2199-2205.

[10] J. AVRON, R. SEILER, Quantization of the Hall conductance for general multiparticle Schrœdinger hamiltonians, *Phys. Rev. Letters*, 54, (1985), 259–262.

[11] J. AVRON, R. SEILER, L. YAFFE, Adiabatic theorems and applications to the quantum Hall effect, *Commun. Math. Phys.*, 110, (1987), 33-49.

[12] M.Ya. AZBELL, Energy Spectrum of a Conduction Electron in a Magnetic Field, *Sov. Phys. JETP*, 19, (1964), 634-645. See also *Phys. Rev. Letters*, 43, (1979), 1954.

[13] A. BEHROOZ, M. BURNS, H. DECKMAN, D. LEVINE, B. WHITEHEAD, P.M. CHAIKIN, Flux Quantization on Quasicrystalline Networks, *Phys. Rev. Letters*, 57, (1986), 368-371.

[14] J. BELLISSARD , Schrödinger operators with an Almost Periodic Potential, in *Mathematical Problems inTheoretical Physics*, R. Schrader, R.Seiler Eds., *Lecture Note in Physics*, 153, (1982), 356-359, Springer Verlag, Berlin–Heidelberg–New-York.

[15] J. BELLISSARD, B. SIMON, Cantor spectrum for the Almost Mathieu equation, *J. Func. Anal.*, 48, (1982), 408–419.

[16] J. BELLISSARD, K–Theory of C*–Algebras in Solid State Physics, *Statistical Mechanics and Field theory, Mathematical aspects*, T.C. Dorlas, M.N. Hugenholtz, M. Winnink Eds., *Lecture Note in Physics*, 257, (1986), 99–156

[17] J. BELLISSARD, D.R. GREMPEL, F. MARTINELLI, E. SCOPPOLA, Localization of electrons with spin-orbit or magnetic interactions in a two dimensional crystal, to appear in *Phys. Rev.*, B33, (1986), 641-644.

[18] J. BELLISSARD, Ordinary Quantum Hall Effect and Non Commutative Cohomology, *To appear in the Proc. of the Bad Schandau Conference on Localization*, Teubner Leipzig, (1987).

[19] G. BERGMANN, Weak localization in thin films, a time-of-flight experiment with conduction electrons, *Phys. Report*, 107, (1984), 1-58.

[20] M. BERRY, Quantal phase factors accompanying adiabatic changes, *Proc. Roy. Soc. London*, A392, (1984), 45-57.

[21] E.I. BLOUNT, Bloch Electrons in a Magnetic Field, *Phys. Rev.*, 126, (1962), 1636-53.

[22] P. BRIET, J.M. COMBES, P. DUCLOS, On the Location of Resonnances for Schrödinger Operators in the Semiclassical Limit II, *Comm. in P. D. E.*, 12(2), (1987), 201-222.

[23] E. BROWN, Bloch Electrons in a Uniform Magnetic Field, *Phys. Rev.*, 133, (1964), A1038-44.

[24] W.G. CHAMBERS, Linear Network Model for Magnetic Breakdown in Two Dimensions, *Phys. Rev.*, A140, (1965), 135-143.

[25] F.H. CLARO, W.H. WANNIER, Magnetic Subband Structure of Electrons in Hexagonal Lattices, *Phys. Rev.*, B19, (1979), 6068-74.

[26] (i)-F.H. CLARO, W.H. WANNIER, Closure of Bands for Bloch Electrons in a Magnetic Field, *Phys. Stat. Sol.*, B88, (1978), K147-151.(ii)-F.H. CLARO, Spectrum of Tight Binding Electrons in a square Lattice with Magnetic Field, *Phys. Stat. Sol.*, B104, (1981), K31-K34.

[27] J.M. COMBES, P. BRIET, P. DUCLOS, Spectral properties of Schrödinger Operators with Trapping Potentials in the Semiclassical Limit, *To appear in the Proc. of the Inter. Conf. on Differential Equations and Mathematical Physics, Birmingham (Alabama), 1986, Springer Verlag* (1987).

[28] A. CONNES, A survey of foliation and operator algebras, *Operator Algebras and applications, D.Kadison Ed.*, 38, Part I&II, (1982), *A.M.S., Providence, Rhode Island.*

[29] A. CONNES, Non Commutative Differential Geometry, *Pub. IHES*, 62, (1986), 43-144.

[30] A. CONNES, Non Commutative Riemanian Manifolds, *Seminar given at the I.H.E.S., May 1987.*

[31] J. DIXMIER, Les C*Algèbres et leurs représentations, *Gauthiers-Villars*, Paris (1969).

[32] B. DOUCOT, R. RAMMAL, Quantum Oscillations in Normal-Metal Networks, *Phys. Rev. Letters*, 55, (1985), 1148-1151.

[33] B. DOUCOT, R. RAMMAL, Interference effects and magnetoresistance oscillations in normal- metal Networks: 1-weak localization approach, *J. Physique*, 47, (1986), 973-999.

[34] B. DOUCOT, W. WANG, J. CHAUSSY, B. PANNETIER, R. RAMMAL, A. VAREILLE, D. HENRY, First Observation of the universal periodic corrections to Scaling : Magnetoresistance of Normal-Metal Self-Similar Networks, *Phys. Rev. Letters*, 57, (1986), 1235-1238.

[35] D.A. DUBROVIN, V.B. MATVEEV, S.P. NOVIKOV, Non linear equations of KdV type, finite zone linear operators and Abelian varieties, *Russ. Math. Surveys*, 31, (1976), 59-146.

[36] M. DUNEAU, A. KATZ, Quasiperiodic Patterns, *Phys. Rev. Letters*, 54, (1985), 2688-2691.

[37] G. ELLIOTT, Gaps in the spectrum of an almost periodic Schrödinger operator, *C.R. Math. Ref. Acad. Sci. Canada*, 4, (1982), 255-259.

[38] H. FESHBACH, Unified Theory of Nuclear Reactions, *Ann. Phys.*, 5, (1958), 357-390 (especially p.363). Unified Theory of Nuclear Reactions II, 19, (1962), 287-313 (formula 2.10).

[39] P. GANDIT, J. CHAUSSY, B. PANNETIER, A. VAREILLE, A. TISSIER, Measurement of the Derivatives of the Magnetization of a Superconducting Networks, *Eur. Phys. Letters*, 3, (1987), 623-628.

[40] P.G. de GENNES, 1)- Diamagnétisme de grains supraconducteurs près d'un seuil de percolation, *C.R. Acad. Sci.*, B292, (1981), 9-12. 2)- Champ critique d'une boucle supraconductrice ramifiée, *C.R. Acad. Sci.*, B292, (1981), 279-282.

[41] P.G. de GENNES, Superconductivity of Metals and Alloys, *Benjamin, New-York,* (1966).

[42] E. GERLACH, D. LANGBEIN, Tight-Binding Approach of Electrons in a Crystal Potential and an External Magnetic Field, *Phys. Rev.*, 145, (1966), 449-457 .

[43] J.M. GHEZ, W. WANG, R. RAMMAL, B. PANNETIER, J. BELLISSARD, Band Spectrum for an Electron on a Sierpinsky Gasket in a Magnetic Field, *To appear in Solid State Comm.*, (1987).

[44] A. GROSSMAN, Momentum-Like Constants of Motion, *1971 Europhysics Conference on Statistical Mechanics and Field Theory, Haifa,* R.N.SEN & C.WEIL Eds.(1972).

[45] See V. GUDMUNDSSON, R.R. GERHARDTS, Interpretation of experiments implying density of states between Landau levels of a two-dimensional electron gas by a statistical model for inhomogeneity, *Phys. Rev.*, B35, (1987), 8005-8014 and references therein.

[46] E.H. HALL, On a new action of the Magnet on Electric Currents, *Amer. J. Math.*, 2, (1879), 287, and *Phil. Mag.*, 9, (1880), 225

[47] P.R. HALMOS, Lectures on Ergodic Theory, *Chelsea Publishing Company,* New–York, (1956).

[48] B.I. HALPERIN, Quantized Hall conductance, current–carrying edge states, and the existence of extended states in a two dimensional disordered potential, *Phys. Rev.*, B25, (1982), 2185-2190.

[49] P.G. HARPER, Single band Motion of Conduction Electrons in a Uniform Magnetic Field, *Proc.Phys. Soc. London,* A68, (1955), 874.

[50] B. HELLFER, J. SJÖSTRAND, (i)-Multiple Wells in the Semiclassical Limit I, *Comm. in P. D. E.*, 9(4), (1984), 337-408.(ii)- Puits Multiples en Analyse Semiclassique II, *Ann. I.H.P.*, 42, (1985), 127-212.

[51] B. HELLFER, J. SJÖSTRAND, Analyse semi classique pour l'équation de Harper, *séminaire de l'Ecole Polytechnique,* p.XVII.1, Avril 1987.

[52] D. HERBERT, R. JONES, Localized States in Disordered Systems, *J.Phys.*, C4, (1971), 1145

[53] D.R. HOFSTADTER, Energy levels and wave functions of Bloch electrons in a rational or irrational magnetic field, *Phys. Rev.*, B14, (1976), 2239.

[54] R.L. HUDSON, K.R. PARTHASARATHY, Quantum Ito's Formula and Stochastic Evolutions, *Commun. Math. Phys.*, 93, (1984), 301-323.

[55] K. ISHII, Localization of Eigenstates and Transport Phenomena in the One Dimensional Disordered Systems, *Supp. to Progress in Theor. Phys.*, 53, (1973), 77.

[56] J. KINOSHITA, K. INAGAKI, C. YAMANOUCHI, K. YOSHIHIRO, J. WAKABAYASHI, S. KAWAJI, An Abrupt Disappearance of Quantum Hall Effect, Observed in Silicon n-Inversion Layer, *to appear in Proc. of 2nd Intern. Symp. on Foundation of Quantum Mechanics in the Light of New Technology, Sept. 86, Phys. Soc. of Japan,* (1987). I thank Pr. Kawaji for informations on this problem.

[57] K. von KLITZING, G. DORDA, M. PEPPER, Realization of a resistance standard based on fundamental constants, *Phys. Rev. Letters*, 45, (1980), 494-497.

[58] P. KRAMER, R. NERI, On periodic and non periodic space fillings of E^n obtained by projections, *Acta Crystallogr.*, A40, (1984), 580-587.

[59] K. KUBO, Statistical Mechanics, *North Holland, Amsterdam* (1967).

[60] R. KUBO, S.I. MIYAKE, N. NASHITSUME, *Solid State Phys.*, 17, (1965), 269-364.

[61] K. KUCHAR, R. MEISELS, G. WEIMANN and W.SCHLAPP, Microwave Hall conductivity of the two-dimensional electron gas, *Phys. Rev.*, B33, (1986), 2965-67.

[62] H. KUNZ, The Quantum Hall Effect for Electrons in a Random Potential, *Commun. Math. Phys.*, 112, (1987), 121-145.

[63] L. LANDAU, Diamagnetismus der Metalle, *Z. für Phys.*, 64, (1930), 629-637.

[64] D. LANGBEIN, Tight Binding and the Nearly-Free-Electron Approach to Lattice Electrons in External Magnetic Fields, *Phys. Rev.*, 180, (1969), 633-648.

[65] R.B. LAUGHLIN, Quantized Hall conductivity in two dimensions, *Phys. Rev.*, B23, (1981), 5652-5654.

[66] D. LEVINE, P. STEINHARDT, Quasicrystals : a New Class of Ordered Structures, *Phys. Rev. Letters*, 53, (1984), 2477-2480.

[67] J.M. LUTTINGER, The Effect of a Magnetic Field on Electrons in a Periodic Potential, *Phys. Rev.*, 84, (1951), 814-817.

[68] D. MERMIN, N. ASHCROFT, Solid State Physics, *Saunders, Philadelphia, Tokyo*, (1976)

[69] J. MOSER, An example of a Schrödinger Operator with an almost periodic potential and a nowhere dense spectrum, *Commun. Math. Helv.*, 56, (1981), 198.

[70] L. ONSAGER, Interpretation of the de Haas-Van Halphen Effect, *Phil. Mag.*, 43, (1952), 1006-1008.

[71] S. OSTLUNDT, R. PANDIT, Renormalization group analysis of a discrete quasiperiodic Schrödinger equation, *Phys. Rev.*, B29, (1984), 1394-1414.

[72] B. PANNETIER, J. CHAUSSY, R. RAMMAL, Experimental Determination of the (H,T) Phase Diagram of a Superconducting Network, *J. Physique Lettres*, 44, (1983), L853-L858.

[73] B. PANNETIER, J. CHAUSSY, R. RAMMAL, J.C. VILLEGIER, Experimental Fine Tuning of the Frustration: 2D Superconducting Network in a Magnetic Field, *Phys. Rev. Letters*, 53, (1984), 1845-1848.

[74] L.A. PASTUR, Spectral Properties of Disordered System in One Body Approximation, *Comm. Math. Phys.*, 75, (1980), 179-196.

[75] G. PEDERSEN, C*–Algebras and their automorphism groups, *Academic Press*, London, New–York, (1979).

[76] R. PEIERLS, Zur Theorie des Diamagnetismus von Leitungelektronen, *Z. für Phys.*, 80, (1933), 763-791.

[77] M. PIMSNER, D. VOICULESCU, Imbedding the irrational rotation C*–Algebra into an AF–Algebra, *J. Operator Theory*, 4, (1980), 211–218.

[78] A.B. PIPPARD, Quantization of coupled orbits in metals, *Proc. Roy. Soc. London*, A270, (1962), 1-13; *Phil. Trans. Proc. Roy. Soc. London*, A256, (1964), 317.

[79] R. PRANGE, Quantized Hall resistance and the measurement of the fine structure constant, *Phys. Rev.*, B23, (1981), 4802–4805.

[80] R. RAMMAL, T.C. LUBENSKY, G. TOULOUSE, Superconducting networks in a magnetic field, *Phys. Rev.*, B27, (1983), 2820-29.

[81] R. RAMMAL, Landau Level spectrum of Bloch electrons in a honeycomb lattice, *J. de Physique*, 46, (1985), 1345-1354.

[82] M.R. RIEFFEL, Irrational rotation C*–algebra, *Short communication to the Congress of Mathematician*, (1978).

[83] D. SCHECHTMAN, I. BLECH, D. GRATIAS, J.V. CAHN, Metallic Phase with Long-Range Orientational Order and No Translational Symetry, *Phys. Rev. Letters*, 53, (1984), 1951-1953.

[84] L. SCHWEITZER, B. KRAMERS, A. Mac KINNON, Magnetic field and electron states in two–dimensional disordered systems, *J. Phys.*, C17, (1984), 4111–4125.

[85] B. SIMON, Trace ideals and their applications, *London Math. Soc. Lecture Notes* 35, Cambridge Univ. Press, (1979).

[86] B. SIMON, Almost Periodic Schrödinger Operators, *Adv. Appl. Math.*, 3, (1982), 463–490.

[87] B. SIMON, Holonomy, the Quantum Adiabatic Theorem and Berry's Phase, *Phys. Rev. Letters*, 51, (1983), 2167-70.

[88] J.B. SOKOLOFF, Unusual band structure, wave functions and electrical conductance in crystals with incommesurable periodic potentials, *Phys. Reports*, 126, (1985), 189-244.

[89] P. STREDA, Theory of quantised Hall conductivity in two dimensions, *J. Phys. C, Solid state*, 15, (1982), L717-L721.

[90] I.M. SUSLOV, Localization in One Dimensional Incommensurate Systems, *Sov.Phys. JETP*,, 56, (1982), 612–617.

[91] See for instance T. TAKAMASU, S. KOMIYAMA, S. HIYAMIZU, S. SASA, Effect of finite electric field on the quantum Hall effect, *Surf. Sci.*, 170, (1986), 202-208 ; Current instability in the quantum Hall Effect, *in Proc. of the ICPS 11-15 August 1986, Stockholm.*, and references therein. I would like to thank Prof. Komiyama for these informations.

[92] C. TANG, M. KOHMOTO, Global scaling properties of the spectrum for a quasiperiodic Schrödinger equation, Phys. Rev., B34, (1986), 2041-2044.

[93] D. THOULESS, A relation between the Density of States and Range of Localization for One Dimensional Random Systems, *J. Phys.*, C5, (1972), 77-81.

[94] D. THOULESS, Localization and the two–dimensional Hall effect, *J. Phys.*, C14, (1982), 3475–3480.

[95] D. THOULESS, M. KOHMOTO, M. NIGHTINGALE, M.den NIJS, Quantized Hall conductance in two dimensional periodic potential, *Phys. Rev. Letters*, 49, (1982), 405-408.

[96] D.J. THOULESS, Bandwidths for a quasiperiodic tight-binding model, *Phys. Rev.*, B28, (1983), 4272-76.

[97] D.C. TSUI, H.L. STORMER, A.C. GOSSARD, Two-Dimensional Magnetotransport in the Extreme Quantum Limit, *Phys. Rev. Letters*, 48, (1982), 1559-62.

[98] D.C. TSUI, H.L. STORMER, A.C. GOSSARD, Zero resistance state of two-dimensional electrons in a quantizing magnetic field, *Phys. Rev.*, B25, (1982), 1405-07.

[99] W. WANG, B. PANNETIER, R. RAMMAL, Thermodynamical description of Berry's topological phase. Magnetization of superconducting networks, *Preprint Grenoble*, (Fev. 1987).

[100] W. WANG, B. DOUCOT, R. RAMMAL, B. PANNETIER, Finite Width Effect on the upper critical Field Hc_2 of Superconducting Networks, *Preprint Grenoble*, (1986).

[101] W. WANG, B. PANNETIER, R. RAMMAL, Quasiclassical Approximations for the Almost Mathieu Equations, *To appear in J. de Physique*, (1987).

[102] C.H. WILCOX, Theory of Bloch Waves, *J. Analyse Math.*, 33, (1978), 146-167.

[103] M. WILKINSON, An example of phase holonomy in WKB theory, *J. Phys.*, A17, (1984), 3459-76.

[104] M. WILKINSON, Critical Properties of Electron Eigenstates in Incommensurate Systems, *Proc. Roy. Soc. London*, A391, (1984), 305-350.

[105] J. ZAK, Magnetic Translation Group, *Phys. Rev.*, A134, (1964), 1602–1607 ; Magnetic Translation Group II: Irreducible Representations, *Phys. Rev.*, A134, (1964), 1607–1611.

[106] I thank Prof. Hajdu for a discussion about it.

[107] J. BELLISSARD, work in preparation.

[108] I thank Prof. Tomyama for comments and informations on that point.

[109] J.RENAULT, A Groupoid approach to C*Algebras, *Lecture Notes in Math.*, 793, (1980), *Springer Verlag, Berlin, Heidelberg, New-York*.

SPIN GROUPS, INFINITE DIMENSIONAL CLIFFORD ALGEBRAS AND APPLICATIONS

A. L. Carey
Centre for Mathematical Analysis, Australian National University,
GPO Box 4, Canberra ACT 2601, AUSTRALIA.

CONTENTS

§1 THE INFINITE DIMENSIONAL COMPLEX ORTHOGONAL AND SPIN GROUPS

The overall reference for this material is a preprint (with title "The infinite complex spin groups") by John Palmer and myself. Background on the general area is provided by the articles of Araki [1], Ruijsenaars [25,26] (note that Araki refers to the Clifford algebra as a 'self–dual CAR algebra') and Palmer [13] although I will try to make this exposition self–contained except for proofs. An interesting reference containing some of the results described here is [34] chapter 12 (which I first read after writing the draft for these notes).

The work discussed here is all inspired by the papers of the Kyoto School [31]. From a mathematical viewpoint the main difficulty with their work is the free use, for infinite dimensional spaces, of results whose proofs are only established in the finite dimensional case. In the latter the methods of proof are purely algebraic and may be found in the papers cited earlier in this paragraph. The generalisation of these finite dimensional results is of interest in its own right, independent of any desire to make rigorous the work of the Kyoto School. In fact it offers the prospect of going beyond their results within the purely infinite dimensional framework.

1.1 Basic facts on Clifford algebras and their representations

We begin with a separable complex Hilbert Space W. For finite dimensional W there are no analytic complications and all that we have to say below is described in Sato, Miwa and Jimbo (henceforth S.M.J.) [31,I]. It is also well known in part in the mathematical literature [4] (cf. also [34]).

Suppose W is equipped with a distinguished symmetric bilinear form (,) which is expressible in terms of the inner product $<\, ,\, >$ via

$$(.,.) = <P.,.>$$

where $P{:}W \to W$ is a fixed complex conjugation. The algebraic Clifford algebra $\mathscr{C}_0(W)$ over W is the algebra with identity I generated by the elements of W where the multiplication satisfies

$$w_1 w_2 + w_2 w_1 = (w_1, w_2)I;\ w_1, w_2 \in W$$

Now $\mathscr{C}_0(W)$ may be completed in an appropriate topology to form a C^*–algebra $\mathscr{C}(W)$ but we will not need this fact explicitly here. Associated with every splitting $W = W_+ \oplus W_-$ into Hilbert subspaces $W_\pm$ isotropic with respect to (,) there is a representation of $\mathscr{C}(W)$. We parametrise such splittings by self adjoint operators $Q : W \to W$ with $Q^2 = I$ where $W_\pm$ are the ± 1 eigenspaces of Q (so $Q = Q_+ - Q_-$ in terms of its spectral projections $Q_\pm$ onto $W_\pm$). This (Fock) representation F_Q is constructed as follows. Introduce the alternating tensor algebra $A(W_+)$ over W_+:

$$A(W_+) = \mathbb{C} \oplus \sum_{k=1}^{\infty} \Lambda^k W_+$$

where $\Lambda^k W_+ = k^{th}$ exterior power. Define for each $x \in W_+, C(x): A(W_+) \to A(W_+)$ by $C(x)v = x \wedge v,\ v \in A(W_+)$. Equipping $A(W_+)$ with the natural pre–Hilbert space structure we complete it (and use the same symbol for the completion) and define

$$F_Q(w) = C(w_+) + C(Pw_-)^*$$

where $w = w_+ + w_- \in W_+ \oplus W_-$ and the $*$ denotes the Hilbert space adjoint.

Any bounded linear transformation G on W satisfying $G^\tau \overset{\text{def}}{=\!=} PG^*P = G^{-1}$ defines an automorphism of $\mathscr{C}(W)$ via

$$w \to Gw, w \in W\ .$$

The set of all such G's forms the complex orthogonal group. We let

$$O_{res}(W) = \{G \mid G \text{ orthogonal, } GQ - QG \text{ is Hilbert–Schmidt}\}\ .$$

If $G \in O_{res}(W)$ we write $G = \begin{bmatrix} A(G) & B(G) \\ C(G) & D(G) \end{bmatrix}$, a 2×2 matrix of operators defined by the splitting $W = W_+ \oplus W_-$ and note that it is not difficult to prove that the off diagonal operators in G are Hilbert–Schmidt while the diagonal ones are Fredholm of index zero.

The subgroup of $O_{res}(W)$ consisting of G satisfying $G^* = G^{-1}$, or equivalently $PGP = G$ is well–studied and goes back to Shale and Stinespring [35]. It is well known that for such G there exists a unitary operator $\Gamma_Q(G)$ on $A(W_+)$ such that

$$\Gamma_Q(G)F_Q(w)\Gamma_Q(G)^{-1} = F_Q(Gw),\ w \in W$$

and that $QG-GQ$ being Hilbert–Schmidt is the appropriate necessary and sufficient condition for the existence of $\Gamma_Q(G)$, [35].

We are interested in part in the generalisation of this result to $O_{res}(W)$. There are numerous reasons for this interest and we list a few here.

(i) Segal and Wilson used a subgroup of $O_{res}(W)$ in their study of the KdV equation [32]. The work of Date et al [11] on the Landau–Lifshitz equation (which we mention briefly in the applications section) suggests that the methods of Segal and Wilson can be extended to the Landau–Lifshitz equation using $O_{res}(W)$.

(ii) It turns out that exactly solvable models in two dimensional quantum field theory are closely connected with the representations of $O_{res}(W)$ and its sub–groups. For example: The massless Thirring model [9], the Federbush model [28], [29], the Luttinger model [5], the continuum limit of the Ising model [17] and the monodromy fields of Palmer [16], [17]. In particular these last two applications require results which utilise infinite dimensional analogues of the SMJ analysis and hence are not covered by the established lore on 'Bogoliubov transformations'.

(iii) Related to (i) and (ii) is the representation theory of loop groups, vertex operators and string theory ([8], [32], [34] and references therein). It is possible that some new ideas will emerge from studying these in the context of $O_{res}(W)$.

1.2 Structure of $O_{res}(W)$:

Now $O_{res}(W)$ has a natural topology arising from using either the norm or strong topology on the 'diagonal' components and the Hilbert–Schmidt topology on the 'off–diagonal' components in the representation $G = \begin{bmatrix} A(G) & B(G) \\ C(G) & D(G) \end{bmatrix}$ determined by the splitting $W = W_+ \oplus W_-$.

In this topology the map $G \to \dim \ker D(G) \pmod 2$ is a continuous homomorphism from $O_{res}(W)$ to $\mathbb{Z}_2$ whose kernel is the connected component of the identity of $O_{res}(W)$. We let $SO_{res}(W)$ denote the latter subgroup.

At first sight it would seem that a subgroup of relevance here would be that generated by those G for which there existed a $g \in \mathscr{C}(W)$ with $gwg^{-1} = Gw$. This is the group introduced for finite dimensional W in SMJ [31,I] and called the Clifford group there. In the infinite dimensional case this requires either $G+1$ or $G-1$ to be trace class (cf. Araki [1]) and this group is thus too small to be of much use. It turns out to be of more interest to consider a slightly larger group

$$SO_Q(W) = \{G \in SO_{res}(W) \mid A(G) \oplus D(G) - I \text{ is trace class}\} .$$

This latter group could be loosely described as the group of Q–inner automorphisms. It should become clearer later why this is the appropriate object to study. More motivation comes from some of the applications [16–24].

A basic fact about $SO_Q(W)$ is the existence of the decomposition

$$SO_Q(W) = SO_{\mathbb{R},Q}(W) B_Q$$

where $B_Q = \{G \in SO_Q(W) \mid G = \begin{bmatrix} * & 0 \\ * & * \end{bmatrix}\}$ and

$$SO_{\mathbb{R},Q}(W) = \{G \in SO_Q(W) \mid PGP = G\}$$

The group B_Q is the analogue of a Borel subgroup of $O(n,\mathbb{C})$ and in fact one can view the Fock representation as defining a line bundle over $SO_Q(W)/B_Q$. I will not develop this Borel–Weil analogy here (however see [34]).

1.3 The Clifford Group

When W is finite dimensional SMJ [31,I] introduced the group $\mathcal{G}(W)$ of invertible elements of $\mathcal{C}(W)$ with

$$gwg^{-1} = G\cdot w,\ w\in W$$

for some othogonal G on W. When W is infinite dimensional we can similarly define $\mathcal{G}_0(W)$ as the group of invertible elements of $\mathcal{C}_0(W)$ with

$$gwg^{-1} = G\cdot w,\ w\in W$$

for some complex orthogonal G.

There is a homomorphism T from $\mathcal{G}_0(W)$ to the complex orthogonal group given by choosing $T_{(g)}$ in the following way for $g \in \mathcal{G}_0(W)$:

$$\begin{aligned} gwg^{-1} &= T(g)w \text{ if } g \text{ is even} \\ gwg^{-1} &= -T(g)w \text{ if } g \text{ is odd} \end{aligned}$$

(note that elements of $\mathcal{C}_0(W)$ are even or odd depending on whether they are in the 1 or -1 eigenspace of the involution on $\mathcal{C}_0(W)$ defined by the map $w\mapsto -w$, $w\in W$).

Now, each $w\in W$ is odd and from the formula

$$wxw^{-1} = 2\frac{(x,w)}{(w,w)}w - x,\ x\in W$$

it is clear that w induces the reflection in the hyperplane orthogonal to w. As every element of $\mathcal{G}_0(W)$ may be written as a product of reflections [3] there is a naturally defined involution τ on $\mathcal{G}_0(W)$ given by

$$g = w_1 \ldots w_k \rightarrow w_k \ldots w_1 = g^{\tau}$$

so that if $g \in \mathscr{G}_0(W)$, $g^{\tau} g \in \mathbb{C}^*$ and the map

$$g \rightarrow g^{\tau} g = \mathrm{nr}(g)$$

is a homomorphism. If W is finite dimensional its kernel is called $\mathrm{Pin}_{\mathbb{C}}(W)$ and there is an exact sequence

$$I \rightarrow \mathbb{Z}_2 \rightarrow \mathrm{Pin}_{\mathbb{C}}(W) \xrightarrow{T} O_{\mathbb{C}}(W) \rightarrow I$$

which restricts on the connected component of the identity $SO_{\mathbb{C}}(W)$ to give

$$I \rightarrow \mathbb{Z}_2 \rightarrow \mathrm{Spin}_{\mathbb{C}}(W) \xrightarrow{T} SO_{\mathbb{C}}(W) \rightarrow I$$

It turns out that these facts have infinite dimensional analogues with $SO_{\mathbb{C}}(W)$ replaced by $SO_Q(W)$ and $\mathrm{Spin}_{\mathbb{C}}(W)$ replaced by a group $\mathrm{Spin}_Q(W)$ whose definition is not immediately obvious, but whose construction is one of the main objectives ofthe preprint of Palmer and myself.

The construction of $\mathrm{Spin}_Q(W)$ does not, on its own, solve the problem of providing the missing analytic details in SMJ [31]. They consider examples of complex orthogonals G which are not in $O_{res}(W)$ and hence for which $\Gamma_Q(G)$ simply does not exist as an operator. However, analogies with the vertex operator construction [6], [32], [34], suggests that one should be able to approximate the SMJ "operators" by genuine operators $\Gamma_Q(G')$ with G' in $O_{res}(W)$ in such a way as to explain what $\Gamma_Q(G)$ really means. This has in fact been done for some of the examples considered by the Kyoto school (for example, the massless Thirring model [9]). The idea of introducing $O_{res}(W)$ is simply that it is a bigger group than those groups of Bogoliubov transformations studied previously and so provides more 'room' in which to look for approximations to the SMJ "operators".

Now let W be infinite dimensional and note that the map T defined above has the property that for $g \in \mathscr{C}_0(W)$, $T(g)$ is in the group

$$O_0(W) = \{G \mid G \text{ complex orthogonal and } G-I \text{ is finite rank}\}$$

If we introduce the involution τ and homomorphism nr on $\mathscr{G}_0(W)$ as in the finite dimensional case and then look in the kernel of nr for those $g \in \mathscr{G}_0(W)$ with $T(g) \in SO_0(W)$ $(=O_0(W) \cap SO_{res}(W))$ we obtain a subgroup denoted $Spin_0(W) \subseteq G_0(W)$,
i.e. $Spin_0(W) = \{g \in \mathscr{G}_0(W) \mid T(g) \in SO_0(W), \ nr(g) = 1\}$.
There is then an exact sequence

$$I \to \mathbb{Z}_2 \to Spin_0(W) \overset{T}{\to} SO_0(W) \to I .$$

This is as far as one can go with algebraic methods.

I can now describe $Spin_Q(W)$, as loosely speaking, the closure of $Spin_0(W)$ in an appropriate topology. To explain how this works in more detail I need some technicalities.

1.4 Preliminaries on Fock states

Let $\Omega_Q = 1 \oplus 0 \oplus 0 \ldots$ denote the "vacuum" state in $A(W_+)$. Define for $g \in \mathscr{G}_0(W)$

$$\langle g \rangle_Q = \langle \Omega_Q, F_Q(g) \Omega_Q \rangle$$

In general we are interested (following SMJ [31]) in the "correlations"

$$\langle g_1 \cdots g_n \rangle_Q = \langle \Omega_Q, F_Q(g_1) \ldots F_Q(g_n) \Omega_Q \rangle$$

and these may be expressed in terms of the $\langle g_i \rangle_Q$ $i=1,\ldots,n$ (see section 2) so that it is of interest to give formulae for the latter. In fact these formulae (some of which are given below) are one of the main reasons for the route we have chosen to $Spin_Q(W)$. If one were not interested in calculating correlations one could follow the more direct argument in [34]. However we have not found a direct route from [34] to the formulae below.

We introduce a 'Wick–ordering' map θ_Q from the algebraic exterior algebra $A(W)$ to $\mathscr{C}_0(W)$ as the unique map satisfying

(i) $\theta_Q 1 = 1$ $(1 = 1 \oplus 0 \oplus \ldots)$

(ii) $\theta_Q(A \wedge B) = \theta_Q(A)\theta_Q(B)$ if $A \in A(W_+)$ or $B \in A(W_-)$.

(cf.[12]).

Now for $G \in SO_0(W)$ it is a simple algebraic construction to find $g \in \mathscr{G}_0(W)$ with $gwg^{-1} = Gw$ and hence to define $\Gamma_Q(G) = F_Q(g)$. (This is contained in [31,I] and discussed further in Palmer [13].) To see how this works note firstly that for $G \in SO_0(W)$ we may as well assume W is finite dimensional (since $G = I$ except on a dimensional subspace). Now if $\{e_k\}_{k=1}^n$ is a basis of W and e_k^* is a dual basis under the pairing defined by $(.,.)$ (so that for example if $\|e_k\| = 1$, $e_k^* = Pe_k$ is a suitable choice), we introduce for each P–skew adjoint operator R on W (i.e. $R^\tau = -R$) the element $\Sigma_j Re_j \wedge e_j^*$, of $\Lambda^2(W)$ which we also write as R. These arise naturally from $G \in SO_0(W)$ by setting $R_Q(G) = (G-1)(Q_-G+Q_+)^{-1}$ and we have the

LEMMA 1: ([31], [13]). *For* W *finite dimensional and* $g \in \mathscr{C}(W)$ *then*

$$g = \langle g \rangle_Q \, \theta_Q(\exp \tfrac{1}{2}(R_Q T(g)) \tag{1.1}$$

Here the exponential is calculated in the exterior algebra. Note that given $G = T(g)$ (1.1) defines g provided we have a formula for $\langle g \rangle_Q$ in terms of G. Note also that when (Q_-G+Q_+) is not invertible this formula fails. There is an analogue [13] however, which takes care of $\ker(Q_-G+Q_+)$. We will not use it here.

Now we are ready to state

LEMMA 2:

$$<g>_Q^2 = \mathrm{nr}(g)\det(Q_-T(g)+Q_+) \tag{1.2}$$

Thus for $\mathrm{nr}(g) = 1$ lemma 1 and lemma 2 determine g in terms of $G = T(g)$ up to sign. The latter ambiguity is inevitable since we are dealing here with a double cover. I believe that (1.2) is related to the Pfaffian formulae in [34] and if so this would provide the link between [34] and the preprint of Palmer and myself.

Now (1.1) and (1.2) hold when W is infinite dimensional for $g \in \mathscr{C}_0(W)$. One simply works inside an appropriate finite dimensional P and Q invariant subspace of W. Thus for $G \in SO_0(W)$ we may use lemmas 1 and 2 to construct $g \in \mathscr{G}_0(W)$ and hence define

$$\Gamma_Q(G) = F_Q(g)$$

Note that since $P{:}W \to W$ extends to define an involution on $\mathscr{C}_0(W)$ written $g \to \bar{g}$, $g \in \mathscr{C}(W)$ we can define $g^* = \bar{g}^T$ and it follows that $F_Q(g)^* = F_Q(g^*)$ so that we have a *–representation of $\mathscr{G}_0(W)$.

1.5 Constructing the infinite dimensional spin group

In closing up $\mathrm{Spin}_0(W)$ in an appropriate topology the only real difficulty is to control which sheet a sequence is converging on. The topology on $\mathrm{Spin}_0(W)$ is nearly fixed by the covering map T, for $SO_0(W) \subseteq SO_Q(W)$ comes equipped with a natural topology whichfor the decomposition

$$G = \begin{bmatrix} A(G) & B(G) \\ C(G) & D(G) \end{bmatrix},$$

amounts to imposing the trace norm topology on the diagonal components and the Hilbert–Schmidt topology on the off–diagonal components.

To control which sheet we are on we use the *–representation F_Q. Thus we introduce a metric ρ on $\mathrm{Spin}_0(W)$ given by:

$$\rho(g_1,g_2) = \tilde{\rho}(T(g_1),T(g_2)) + \|F_Q(g_1)\Omega_Q - F_Q(g_2)\Omega_Q\|$$

where $\tilde{\rho}$ is the metric on $SO_o(W)$. Then $Spin_Q(W)$ is just the closure of $Spin_o(W)$ in this topology. The advantage of this definition is that in order to prove that it works one needs to construct simultaneously, an action of $Spin_Q(W)$ on the Fock space $A(W_+)$ by densely defined, invertible, but unbounded operators. It is this last fact which makes the introduction of some analysis necessary. A second problem is that to analyse ρ we need to consider $<g_n>_Q$, for sequences $\{g_n\}_{n=1}^{\infty}$ even when the sequence $<g_n>_Q$ converges to zero. Now we have skirted the possibility that $<g>_Q = 0$ or equivalently that $\ker(Q_-T(g)+Q_+) \neq 0$. In fact of course the set of $G \in SO_{res}(W)$ with $\ker(Q_-G+Q_+) = 0$ (i.e. $<g>_Q \neq 0$) is just an open subset and we need a technique for handling G outside this subset. I do not want to go into how this problem is solved here except to say that it makes the constru tion of $Spin_Q(W)$ by this method rather complicated in its details.

The upshot of this then is the existence of a group $Spin_Q(W)$ and an exact sequence

$$I \to \mathbb{Z}_2 \to Spin_Q(W) \xrightarrow{T} SO_Q(W) \to I$$

where the covering map T extends that defined previously. Now a corollary of this is the existence of a representation of $Spin_Q(W)$ on $A(W_+)$. If we let θ denote the closure of the orbit of 1_Q under the action of $Spin_o(W)$ and let $\mathscr{D}$ denote the dense linear subspace $F_Q(\mathscr{C}_o(W))\theta$ of $A(W_+)$ then there is a strongly continuous representation Γ_Q of $Spin_Q(W)$ on $\mathscr{D}$ (by operators which are unbounded in general) such that

$$\Gamma_Q(g)F_Q(w)\Gamma_Q(g)^{-1} = F_Q(T(g)w)$$

where the equality holds only on $\mathscr{D}$.

Now we come finally to our main result. To obtain $SO_{res}(W)$ from $SO_Q(W)$

we need to include the group of all complex orthogonals on W which have the form $\begin{bmatrix} G_0 & 0 \\ 0 & (G_0^\tau)^{-1} \end{bmatrix}$ relative to the decomposition $W=W_+ \oplus W_-$. Here $G_0{}^\tau$ denotes the transpose of $G_0 : W_+ \to W_+$ relative to $(.,.)$. This latter group we denote $GL(W_+)$ and form the semidirect product

$$\mathrm{Spin}_Q(W) x_\alpha \, GL(W_+)$$

where x_α denotes the fact that $GL(W_+)$ acts by conjugation on $\mathrm{Spin}_Q(W)$ (one needs to check that this action is well defined). This group acts on $\mathscr{D}$ as follows. Define $\Gamma(G_0) = 1 \oplus G_0 \oplus G_0 \otimes G_0 \oplus \ldots$ and hence an action of G_0 on $F_Q(\mathscr{C}_0(W))1_Q$. Extend this to all of $\mathscr{D}$ by writing

$$\Gamma(G_0)\Gamma_Q(g)1_Q = \Gamma_Q(G_0 g G_0^{-1})\Gamma(G_0)1_Q \, .$$

Then the representation of $SO_{res}(W)$, $\Gamma_Q x\Gamma$, defined by

$$\Gamma_Q x\Gamma(g,G_0) = \Gamma_Q(g))\Gamma(G_0)$$

factored by its kernel gives a group which we denote $\mathfrak{S}\mathrm{pin}_Q(W)$. One then has the exact sequence

$$I \to \mathbb{C}^* \to \mathfrak{S}\mathrm{pin}_Q(W) \overset{T}{\to} SO_{res}(W) \to I \tag{1.3}$$

Clearly $\mathfrak{S}\mathrm{pin}_Q(W)$ acts by operators on $\mathscr{D}$ which implement the automorphisms of $\mathscr{C}(W)$ defined by the corresponding elements of $SO_{res}(W)$. To get $O_{res}(W)$ we need only give the action of the reflections defined by $w \in \mathscr{C}_0(W)$ and this is easily done.

From the viewpoint of the Kyoto School the significance of the exact sequence (1.3) is indicated in T. Miwa "An introduction to the theory of τ–functions" Springer Lecture Notes in Physics Vol.242 pp.96–142.

Now (1.3) is pretty well the best one can do in the direction of producing infinite dimensional analogues of the finite dimensional results in [31]. To go further and use them in a direct way to make rigorous the applications in [31] requires further analysis.

§2 APPLICATIONS

2.1 Preliminaries

An account of the underlying general strategy of SMJ may be found in "Monodromy, solitons and infinite dimensional Lie algebras" by M. Jimbo and T. Miwa in Vertex operators in Mathematics and Physics (ed. J. Lepowsky, S. Mandelstam, I.M. Singer, Springer Berlin 1984). Their idea in a nutshell is that particular spherical functions (or limits thereof) for the representation of $O_{res}(W)$ provide solutions of non–linear differential equations including the Painlevé equations (in the case of the scaling limit of the Ising model correlations), the KdV and K.P. heirarchy and the Landau–Lifshitz equation. To describe this work fully is beyond the scope of this lecturer. Instead I wish to indicate how the discussion in the preceding section is relevant to making sense of the "vertex" operators used in [31] and [11].

2.2 Ising model, continuum limit

To save space I will assume that the notion of taking a continuum limit of a lattice theory is understood. In heuristic terms the Ising model case is explained in [31]. A more precise mathematical discussion may be found in [14], [15]. In any case it is generally agreed that in the continuum limit the Ising model becomes a free fermion theory, describable in terms of a Fock representation of the Clifford algebra. The non–triviality of the Ising model arises because one is not interested in the fermions or Clifford algebra elements as such but in the 'spins' (which may be thought of as defining elements of the complex orthogonal group) or more precisely the continuum limit of the spin operators. The latter may be described as 'operators representing' elements of the complex orthogonal group which do not lie in $O_{res}(W)$. Clearly, making sense of what is meant by 'operator representing' is where the difficulties arise.

Since the quantum field theory constructed from the continuum limit of the spin operators has a non–trivial S–matrix [23] it is not surprising that significant mathematical problems need to be solved.

The continuum limit of the Ising model is constructed on the Hilbert space $W = L^2(\mathbb{R},\mathbb{C}^2)$. The complex conjugation P takes the form

$$P = \begin{bmatrix} 0 & \Lambda \\ \Lambda & 0 \end{bmatrix}$$

where $(\Lambda f)^\wedge(p) = \overline{f^\wedge(p)}$ with the Fourier transform $\hat{f}$ of f being defined via $f(x) = \frac{1}{\sqrt{2\pi}}\int_{-\infty}^{\infty} \hat{f}(p)e^{ipx}dp$.

We can now form the Clifford algebra $\mathscr{C}(W)$ and consider the representation given by taking up Q to be multiplication by

$$\begin{bmatrix} -\epsilon(p) & 0 \\ 0 & \epsilon(p) \end{bmatrix}$$

where $\epsilon(p) = \begin{cases} 1 & p>0 \\ -1 & p\leq 0 \end{cases}$. Note that this Q is obtained from SMJ [31,IV] by setting their 'mass parameter' equal to zero. We could consider non–zero 'mass' in what follows equally well, but the formulae are a little more complicated. Note that the 'mass' parameter corresponds to the correlation length. In the Ising model it is still an open problem to determine the correlation functions when the 'mass' parameter is zero. SMJ do not consider this case and the rigorous methods of Palmer do not definitively answer the question. I will describe this problem from the Spin group viewpoint below (cf. also [29]).

To facilitate translation from SMJ to the notation used by Palmer and the present notes I include the following table:

This paper	SMJ
P	J
Q	E
PQ	H
PQ_+	K
R	RJ^{-1}

The operator of interest to us (which will define the continuum limit of the spin operator) is multiplication by Φ_y where

$$\Phi_y f(x) = \begin{bmatrix} \epsilon(x-y) & 0 \\ 0 & \epsilon(x-y) \end{bmatrix} f(x),\ f \in L^2(\mathbb{R},\mathbb{C}^2)\ .$$

It is easy to check that Φ_y is complex orthogonal but does not lie in $O_{res}(W)$. Now for those G in $SO_{res}(W)$ the method employed by SMJ (and in the previous section of course) to calculate $\Gamma_Q(G)$ is to define,

(i) $$R(G) = (G-1)(Q_-G+Q_+)^{-1}$$

(ii) $$\Gamma_Q(G) = \Gamma_Q\left[\langle g\rangle_Q\ \theta_Q(\exp \tfrac{1}{2}R(G))\right]$$

where $T(g) = G$. Now strictly speaking we have only defined $\Gamma_Q\left[\theta_Q(\exp \tfrac{1}{2}R(G))\right]$ for G in $SO_0(W)$ but this definition can be extended, following Palmer [13] to those $R(G)$ for which $Q_-R(G)Q_+ + Q_+R(G)Q_-$ Hilbert–Schmidt.

To assign a meaning to the SMJ "operator" or "field" $\Gamma_Q(\Phi_y)$ (which formally is just a 'vertex operator') one could try to approximate it in some sense by $\Gamma_Q(G)$'s for G in $O_{res}(W)$. To see what we are approximating let us proceed for a moment as in [31, IV]. So first we must calculate $R(\Phi_y)$ and hence we need a method for finding $(Q_-G+Q_+)^{-1}$.

This inverse is computed following SMJ [31,IV] using the

LEMMA *If there exist densely defined operators* $X_{\pm}$ *on* W *with*

$$(G-1)X_{+}(G+1) = 0$$
$$(G+1)X_{-}(G-1) = 0$$

then
$$R(G) = X_{-}^{-1}(G-1)X_{+}$$

and
$$(Q_{+}+Q_{-}G)^{-1} = \tfrac{1}{2}\left[{}_{+}^{-1}(1+G) + X_{-}^{-1}(1-G)\right]X_{+}\,.$$

This lemma is proved in [31,IV] and is just algebraic. Of course it is slightly mathematically imprecise since no domain has been given for $X_{\pm}$ and clearly if the latter are unbounded (as is the case for $\dot{\Phi}_y$) then so is $R(G)$ in general. Note that this corresponds to the fact that zero is in the continuous spectrum of $Q_{+}+Q_{-}G$ and so the lemma is computing the (unbounded) inverse in that case.

We record X_{+} and X_{-} for Φ_y as the respective multiplication operators on the Fourier transform given by the matrix functions

$$\frac{1}{\sqrt{2}}\begin{bmatrix}\eta(p)\,|p|^{-\frac{1}{2}} & 0\\ 0 & \eta(p)\,|p|^{\frac{1}{2}}\end{bmatrix}\begin{bmatrix}1 & 1\\ -1 & 1\end{bmatrix},$$

$$\frac{1}{\sqrt{2}}\begin{bmatrix}0 & \eta(p)\,|p|^{-\frac{1}{2}}\\ \overline{\eta(p)}\,|p|^{\frac{1}{2}} & 0\end{bmatrix}\begin{bmatrix}1 & 1\\ -1 & 1\end{bmatrix},$$

where $\eta(p) = \begin{cases}1 & p>0\\ i & p\leq 0\end{cases}.$

Then one calculates $R_y = (\Phi_y-1)(Q_{-}\Phi_y + Q_{+})^{-1}$ to be the integral operator on the Fourier transform with matrix kernel

$$R(p,q,y) = \frac{1}{2}\,\frac{ie^{-i(p-q)y}}{\pi(p-q-i0)}\begin{bmatrix}1 & -1\\ 1 & 1\end{bmatrix}\begin{bmatrix}0 & \eta(p)\,|p|^{-\frac{1}{2}}\\ \overline{\eta(p)}\,|p|^{\frac{1}{2}} & 0\end{bmatrix}$$
$$\cdot\begin{bmatrix}\eta(q)\,|q|^{-\frac{1}{2}} & 0\\ 0 & \eta(q)\,|q|^{\frac{1}{2}}\end{bmatrix}\begin{bmatrix}1 & 1\\ -1 & 1\end{bmatrix}$$

It is easy to check that $Q_{-}R_yQ_{+}$ is not Hilbert–Schmidt so one can proceed no further along this path except to note, following Ruijsenaars [27,28,29], that the object which SMJ write down formally as

$$\Gamma_Q \theta_Q(\exp \tfrac{1}{2} R_y)$$

can be given a meaning as a densely defined quadratic form on the space $A(W_+)$. One is left then with the technical question of whether one may actually define, via the map

$$f \to \int_{-\infty}^{\infty} \Gamma_Q(\theta_Q(\exp \tfrac{1}{2} R_y))f(y)dy, \ f \in S(\mathbb{R}) ,$$

an operator valued distribution. While this point has not been resolved in any direct way, in the massive case it may be deduced from results in [IX], [15] and [28], although the details of the argument have not been published.

There are at least two possible indirect approaches one of which has been successfully carried out. Following other work in this area [8], [9], [28], [37], [34], one might try to replace R_y by an approximate $R_{y,\epsilon}$ corresponding to some $G_{y,\epsilon}$ in $SO_{res}(W)$, with the property that as $\epsilon \to o$, $R_{y,\epsilon} \to R_y$ in such a way that one could control the convergence of $\Gamma_Q \theta_Q(\exp \tfrac{1}{2} R_{y,\epsilon})$. I have not succeeded in doing this. On the other hand Palmer [16],[17], has shown how to produce approximations to $\Gamma_Q(\theta_Q(\exp \tfrac{1}{2} R_y))$ at least in the case of 'nonzero mass' by going back to the lattice. I will briefly review his work here.

The Clifford algebra approach to the Ising model via the transfer matrix is now well–understood in mathematical terms cf.[2] and references therein. I will not go over old ground here but try to give a brief resumé of Palmer's approach [16], [17]. The Ising model Clifford algebra 'lives on' $H = \ell^2(\mathbb{Z}) \oplus \ell^2(\mathbb{Z})$. It is convenient however to also work with the Fourier transform $L^2(S^1) \oplus L^2(S^1) = L^2(S^1, \mathbb{C}^2)$. Palmer works with the Q–operator given by multiplication by

$$Q(\Phi) = \begin{bmatrix} 0 & ie^{i\alpha(\Phi)} \\ -ie^{-i\alpha(\Phi)} & 0 \end{bmatrix}, \Phi\in[-\pi,\pi]$$

where $\alpha(\Phi)$ is defined (together with the function γ) by the relations

$$\cosh \gamma(\Phi) = \frac{(\cosh 2K^*)^2}{\sinh 2K^*} - \cos\Phi$$

$$e^{i\alpha(\Phi)} \sinh\gamma(\Phi) = \coth 2K^* - \cosh 2K^* \cos\Phi + i \sinh 2K^* \sin\Phi$$

$$\alpha(0) = 1$$

and K^* is a parameter related to the temperature (in this parametrisation the critical temperature corresponds to $\sinh 2K^* = 1$).

The crucial operator in this context is the so–called transfer matrix which acts as multiplication by the matrix function

$$T(\Phi) = \begin{bmatrix} \frac{(\cosh 2K^*)^2}{\sinh 2K} - \cos\Phi & \begin{matrix}\sinh 2K^* \sin\Phi \\ -i(\coth 2K^* - \cosh 2K^* \cos\Phi)\end{matrix} \\ \begin{matrix}\sinh 2K^* \sin\Phi \\ +i(\coth 2K^* - \cosh 2K^* \cos\Phi)\end{matrix} & \frac{(\cosh 2K^*)^2}{\sinh 2K} - \cos\Phi \end{bmatrix}$$

To define the 'discrete' analogue of Φ_y we set

$$f^{\vee}(k) = \frac{1}{\sqrt{2\pi}}\int_{-\pi}^{\pi} f(\Phi)e^{-ik\Phi}d\Phi\ ,\ k\in\mathbb{Z} + \tfrac{1}{2}$$

and define a map on H by

$$(S(\lambda)f)^{\vee}(k) = \tfrac{1}{2}((1-\epsilon(k) + (1+\epsilon(k))\lambda)f^{\vee}(k)\lambda\in\mathbb{C}\ .$$

The operators $s(\lambda)\in O_{res}(H)$ except in special cases. Using translation on $\mathbb{Z} + \frac{1}{2}$ and powers of the transfer matrix one may define for each

$\underline{a} = (a_1,a_2)\in\mathbb{Z}^2$ an operator $s_{\underline{a}}(\lambda)$ on H which for $\lambda = -1$ is exactly the image under T of the Ising model spin operator. When $\lambda = -1$ and

$\sinh 2K^* = 1$, $s_{\underline{a}}(\lambda) \notin O_{res}(H)$, and this is the source of the difficulty in defining the Ising model at criticality and in the scaling limit (i.e. lattice spacing goes to zero). Nevertheless by considering the correlations

$$<\Omega_Q, \Gamma_Q(s_{\underline{a}_1}(-1)) \dots \Gamma_Q(s_{\underline{a}_n}(-1))\Omega_Q> \tag{2.1}$$

Palmer and Tracy have been able to control the limit as $\sinh\ 2K^*$ goes to 1 and the lattice spacing goes to zero in such a way as to give meaning to the correlations of the SMJ "fields" $\Gamma_Q(\Phi_y)$ and hence to the quantum field theory they describe. (This is the 'nonzero mass' or 'massive' scaling limit). If we fix $\sinh 2K^* = 1$ but have $\lambda \neq -1$ then the limit as the lattice spacing goes to zero of the analogue of (2.1) can also be controlled. (This is the critical scaling limit of the monodromy fields $s_{\underline{a}}(\lambda)$.) The limit $\lambda \to -1$ remains a mystery however. (Another approach, not using $0_{res}(W)$, to this question appears in [36]).

It would be more satisfactory if I could report a short description of the techniques involved but unfortunately they are too complicated. Suffice to say that the key technique is the following 'product formula' [16],[13].

$$<\Omega_Q, \Gamma_Q(g_1) \dots \Gamma_Q(g_n)\Omega_Q> = \prod_{k=1}^{n} <g_k> \det_2(1+LR) \tag{2.2}$$

where $\det_2$ is the 'renormalised' determinant of the operator $1+LR$ with LR a Hilbert–Schmidt operator constructed as follows.

Let $T(g_k) = \begin{bmatrix} A_k & B_k \\ C_k & D_k \end{bmatrix}$ be the decomposition of $T(g_k)$ with respect to the splitting $W = W_+ \oplus W_-$. Then if D_k is invertible for all k we may define L to be the $n \times n$ block matrix with entries e_{ik} given by

$$i < k \quad e_{ik} = \begin{cases} -Q_+ & k = i+1 \\ -A_{i+1} Q_+ & k = i+2 \\ -A_{i+1} \cdots A_{k-1} Q_+ & k > i+2 \end{cases}$$

$$i > k \quad e_{ik} = \begin{cases} Q_- & i = k+1 \\ D_{k+1}^{-1} Q_- & i = k+2 \\ D_{i-1}^{-1} \cdots D_{k+1}^{-1} Q_- & i > k+2 \end{cases}$$

$$i = k \quad e_{ii} = 0$$

and R to be the block diagonal matrix with diagonal entries R_k where $k = 1,2,\ldots,n$ and

$$= \begin{bmatrix} -B_k D_k^{-1} C_k & B_k D_k^{-1} \\ D_k^{-1} C_k & 0 \end{bmatrix}$$

Note that this formula fails if the D_k are not invertible. In the case where $g_i = s_{\underline{a}_i}(\lambda)$, $i = 1,\ldots,n$ the RHS of (2.1) may be computed and thus one gains control over the various limits described above.

2.3 Fermions on an elliptic curve

Elliptic functions are ubiquitous in the statistical mechanical solvable model game. The elliptic curves which underly the occurrence of these functions have not seemed so significant. However the parametrisation of solutions of the KdV equation (cf. [33]) by elliptic curve data and the use of a particular elliptic curve to construct solutions of the Landau–Lipshitz equation [11] suggest to me a deeper role. It is also tempting to speculate that the occurrence of Virasoro algebras in connection with statistical mechanics (i.e. so–called conformal models) could be understood by going to an 'elliptic curve picture'. An indication of this is given by the construction of Itoyama and Thacker in a 1986 Fermilab preprint entitled "Lattice Virasoro algebra and corner transfer matrices in the Baxter eight–vertex model". This employs a Clifford algebra $\mathscr{C}(W)$ where W is (an unspecified) space of functions depending on the spectral (elliptic curve) variable which parametrises Baxter's family of commuting transfer matrices. Extensive technical use of the elliptic curve picture defined by the spectral representation of the transfer matrix of the two dimensional Ising model in [31] and [16] is a further indication. Rather than continue this speculation let me introduce instead a somewhat different looking example where the complex orthogonal group enters in an essential way.

In [11] Daté et al consider 'fermions on an elliptic curve' which in our context amounts to introducing a Clifford algebra $\mathscr{C}(W)$ where W is a space of functions on an elliptic curve E. In the notation of [11] this is the curve parametrised by complex parameters ω and k with

$$\omega^2 = (k^2-a^2)(k^2-b^2) \quad a,b\in\mathbb{C}.$$

In this parametrisation there are two points at infinity, ∞_+, ∞_- given by letting $k,\omega \to \infty$ with $\frac{\omega}{k^2} \to \pm 1$. Choose a path γ which encircles ∞_+ and ∞_- once in the clockwise direction.

(Note γ is contractible to a point in E). Let N be a neighbourhood of ∞_+ and ∞_- containing γ and consider functions $f{:}N\backslash\{\infty_+,\infty_-\} \to \mathbb{C}$ which are analytic and extend to meromorphic functions on E. One may show that there is a splitting of the space W_0 of all such functions into subspaces, $W_0 = W_+ \oplus W_-$, specified by the requirement that if $f\in W_0$, $f = f_+ + f_-$, $f_\pm \in W_\pm$, where f_+ is analytic on $E\backslash\{\infty_+,\infty_-)$ and f_- is analytic in $N\backslash\{\infty_-\}$ having at most a simple pole at ∞_- and at least a simple zero at ∞_+.

More generally, given any Riemann surface M of genus g points $\infty_+,\infty_- \in M$ and a neighbourhood N of $\{\infty_+,\infty_-\}$, there is a line bundle L over M with divisor $g\infty_+ - \infty_-$ and a splitting of the space W of holomorphic sections of $L|_{N\backslash\{\infty_+,\infty_-\}}$ into holomorphic sections of $L|_N(\equiv W_+)$ and holomorphic sections of $L|_{E\backslash\{\infty_+,\infty_-\}}(\equiv W_-)$.

There is an involution $\#$ on E with no fixed points. (If E is thought of as $\mathbb{C}/L$ where L is a lattice with periods ω_0,ω_1 then $\#$ amounts to translation by half a period, either $\frac{1}{2}\omega_0$ or $\frac{1}{2}\omega_1$). In the coordinates above $(k,\omega)^\# = (-k,-\omega)$. With $dp = \frac{dk}{2\pi i\,\omega}$ a holomorphic 1–form, we introduce a bilinear form $(\ ,\ )$ on W_0:

$$(f,g) = \int_\gamma f(p^\#)g(p)dp\ , \quad f,g\in W_0 .$$

Then one may show that $(\ ,\ )$ is non–degenerate on W_0 and hence form the

Clifford algebra $\mathscr{C}(W_0)$. The complex conjugation P does not seem to have an intrinsic definition and I will not discuss it here. The splitting $W_0 = W_+ \oplus W_-$ defines a Q operator in the usual way and so there is a natural Fock representation F_Q of $\mathscr{C}(W_0)$.

The subgroup of the complex orthogonal group on W_0 of interest here is the group $\mathscr{G}$ of meromorphic functions Φ on E with $\Phi: N\backslash\{\infty_+,\infty\} \to \mathbb{C}^*$ analytic and such that $\Phi(p^\#) = \frac{1}{\Phi(p)}$. (This is the analogue for E of the loop group of $\mathbb{C}^*$ which acts on $L^2(S^1,\mathbb{C})$ in the more familiar case [34]). One may show that such Φ lie on $O_{res}(W)$. There is a representation of a Heisenberg algebra' on $A(W_+)$ constructed from representing functions in the Lie algebra of $\mathscr{G}$ by operators on $A(W_+)$. Using this Heisenberg algebra Daté et al write down 'τ–functions' which give solutions of the Landau–Lifshitz equation and also fermionic 'vertex operators' which appear to implement a 'boson–fermion' correspondence in this context (cf. [6], [34] for the more usual case).

It would be of interest to work out the details of this, in particular to see how the ideas of Segal–Wilson [33] translate to this context.

Joint work with Keith Hannabuss and Michael Eastwood has elucidated some of the mathematical structure of this example. Here is a brief summary of our results.

(i) The key idea appears to be to use the properties of meromorphic functions on E as a substitute for the analytic machinery of [6], [8], [32], [34], using this it appears likely that generalisations to hyperelliptic curves are possible.

(ii) Each $\Phi \in \mathscr{G}$ has $Q_+\Phi Q_- + Q_-\Phi Q_+$ finite rank. Hence $\Gamma_Q(\Phi)$ can be given fairly explicity.

(iii) We have the following 'index formula'

dim ker $Q_-\Phi Q_-$ = number of poles of Φ in N (mod.2).

(iv) A central extension of $\mathscr{G}$ acts cyclically on the Fock space.

While there appears to be no analogue of the 'blip construction' of [6], [8], [32],

[34], we nevertheless obtain a weak form of the boson–fermion correspondence in the sense that a general Fermion state in $A(W_+)$, say

$$F_Q(w_1)F_Q(w_2)\ldots F_Q(w_n)\Omega_Q\ ,\ w_i \in Q_+w,\ i = 1,2,\ldots,n$$

may be constructed by acting on Ω_Q with operators obtained from certain $\Gamma_Q(\Phi)$ for $\Phi \in \mathcal{G}$.

We intend to report more fully on these results elsewhere.

ACKNOWLEDGEMENT

The ideas of this last section were initiated by conversations with John Palmer whose assistance is gratefully acknowledged. These notes formed the basis of lectures given at Warwick and also at the Centre for Mathematical Analysis at the Australian National University, Canberra.

I would like to record my thanks to Dai Evans for his invitation to participate in the Warwick Symposium on Operator Algebras and Applications and to the members of the Warwick Mathematics Institute for their hospitality during my stay. I would also like to thank the Centre for Mathematical Analysis and all those involved in the program on Harmonic Analysis and Operator Algebras for making my stay in Canberra so enjoyable.

Finally, I would like to thank Simon Ruijsenaars for help in elucidating some finer points.

REFERENCES

[1] Araki, H., "On quasifree states of the CAR and Bogoluibov automorphisms", Publ. Res. Inst. Math. Sci. 6 (1970) 385–442.

[2] Araki, H., Evans, D.E., "On a C* algebra approach to phase transition in the 2 dimensional Ising model", Commun. Math. Phys. 91 (1983) 489–503.

[3] Artin, E., Geometric Algebra, Interscience, New York, (1957).

[4] Atiyah, M.F., Bott, R., Shapiro, A., "Clifford modules", Topology 3 (1964) 3–38.

[5] Carey, A.L., Hannabuss K.C., "Loop groups, theta functions and the Luttinger model", J.Func.Anal. (to appear).

[6] Carey, A.L., Hurst, C.A., "A note on the boson–fermion correspondence and infinite dimensional groups", Commun. Math. Phys. 98 (1985) 435–448.

[7] Carey, A.L., O'Brien, D.M., "Automorphisms of the infinite dimensional Clifford algebra and the Atiyah–Singer mod 2 index", Topology 22 (1983) 937–948.

[8] Carey, A.L., Ruijsenaars, S.N.M., "On Fermion gauge groups, current algebras and Kac–Moody algebras", Acta. App. Math. (to appear).

[9] Carey, A.L., Ruijsenaars, S.N.M., Wright, J.D., "The massless Thirring model: positivity of Klaiber's n–point functions", Commun. Math. Phys. 99 (1985) 347–364.

[10] Daté, E., Jimbo, M., Kashiwara, M., Miwa, T., "Transformation groups for soliton equations": I. Prog. Jpn. Acad. 57A (1981) 342–347; II. Ibid 387–392; III. J. Phys. Soc. Jpn. 50 (1981) 3806–3812; IV. Physica 4D $^{8}{}_{1982}{}^{9}$ 343–365; V. RIMS preprint; VI. J. Phys. Soc. Jap. 50 (1981) 3813–3818; VII. RIMS preprint.

[11] Daté, E., Jimbo, M., Kazshiwara, M., Miwa, T., "Landau–Lifshitz equation: solitons, quasiperiodic solutions and infinite dimensional Lie algebras", J. Phys. A16 (1983) 221–336.

[12] Gross, L., "On the formula of Mathews and Salam", J. Func. Anal. 25 $^{8}{}_{1977}{}^{9}$ 162–209.

[13] Palmer, J., "Products in spin representations", Ad. App. Math. 2 (1981) 290–328.

[14] Palmer, J., Tracy, C., "Two–dimensional Ising correlations: convergence of the scaling limit", Ad. App. Math. 2 (1981) 329–388.

[15] Palmer, J., Tracy, C., "Two dimensional Ising correlations: the S.M.J. analysis", Ad. Appl. Math. 4 (1983) 46–102.

[16] Palmer, J., "Monodromy fields on $\mathbb{Z}^2$ ", Commun. Math. Phys. 102 (1985) 175–206.

[17] Palmer, J., "Critical scaling for monodromy fields", Commun. Math. Phys. 104 (1986) 353–385.

[18] Palmer, J., "A Grassmann calculus for infinite spin groups", preprint (Arizona 1986).

[19] Palmer, J., "A Grassmann calculus for $\hat{GL}(H)$ ", in preparation.

[20] Pickrell, D., "On $U_{(2)}$ invariant measures, preprint.

[21] Pickrell, D., "Measures on infinite dimensional Grassmann manifolds", J. Func. Anal. (to appear).

[22] Pickrell, D., Thesis, Univ. of Arizona, (1984).

[23] Pickrell, D., "On the support of quasi–invariant measures on infinite dimensional Grassmann manifolds", Trans. Amer. Math. Soc. (to appear).

[24] Pickrell, D., "Decomposition of regular representations of $U_2(H)$", Pac. J. Math. (to appear).

[25] Ruijsenaars, S.N.M., "On Bogoliubov transformations", J. Math. Phys. 18 (1977) 517–526.

[26] Ruijsenaars, S.N.M., "On Bogoliubov transformations II: the general case", Ann. Phys. 116 (1978) 105–134.

[27] Ruijsenaars, S.N.M., "Integrable quantum field theories and Bogoliubov transformations", Ann. Phys. 132 (1981) 328–382.

[28] Ruijsenaars, S.N.M., "The Wightman axioms for the fermionic Federbush model", Commun. Math. Phys. 87 (1982) 181–228.

[29] Ruijsenaars, S.N.M., "Scattering theory for the Federbush, massless Thirring and continuum Ising models", J. Func. Anal. 48 (1982) 135–171.

[30] Ruijsenaars, S.N.M., "On the two–point functions of some integrable relativistic quantum field theories", J. Math. Phys. 24 (1983) 922–931.

[31] Sato, M., Miwa, T., Jimbo, M., "Homonomic quantum fields I–V", Publ. Res. Inst. Math. Sci. 14 (1978) 223–267; 15 (1979) 201–278; 15 (1979) 577–629; 15 (1979) 871–972; 16 (1980) 531–584.

[32] Segal, G., "Unitary representations of some infinite dimensional groups" Commun. Math. Phys. 80 (1981) 301–362.

[33] Segal, G., Wilson, G., "Loop groups and equations of KdV type". Publ. IHES. 61 (1985) 5–65.

[34] Segal, G., Pressley, A.N., "Loop groups", Oxford Univ. Press (1986).

[35] Shale, D., Stinespring, W., "States of the Clifford algebra", Ann. of Math. 8 #2 (1964) 365–381.

[36] O'Carroll, M., and Schor, R., "The scaling limit and Osterwalder–Schroder axioms for the two–dimensional Ising model", Commun. Math. Phys. 84 (1982) 153–170.

1 SUBFACTORS AND RELATED TOPICS

V.F.R. Jones
Department of Mathematics, University of California, Berkeley, CA 94720

Given a subfactor N of a II_1 factor M with the same identity, one defines the index of N in M as $[M:N] = \dim_N(L^2(M))$ = the Murray von Neumann coupling constant for N on the Hilbert space $L^2(M)$, (= completion of M with respect to the inner product $\langle a,b\rangle = tr(ab^*)$, tr being the unique normalized trace on M). The following result shows the interest of this notion.

<u>Theorem</u> a) If $[M:N] < 4$ then there is an $n \in \mathbb{Z}$, $n \geq 3$, with $[M:N] = 4\cos^2\pi/n$.

b) For any real $r \geq 4$ there is a pair $N \subseteq M$ with $[M:N] = r$.

The "basic construction" of the theory is as follows. If $N \subseteq M$ are finite von Neumann algebras and tr is a faithful normal normalized trace on M, one considers the von Neumann algebra $\langle M,e_N\rangle = \{M,e_N\}''$ on $L^2(M,tr)$ where e_N is the orthogonal projection onto $L^2(N,tr)$ (tr restricted to N). If N and M are factors then $[M:N] < \infty$ iff $\langle M,e_N\rangle$ is a II_1 factor and then $[M:N]tr(e_N) = 1$.

Ocneanu has made great progress on classifying subfactors of the hyperfinite II_1 factor with given index, which he will explain in his talk. It would appear that the classification is complete for index < 4. Wenzl has constructed many examples of subfactors in index > 4 with the additional property $N' \cap M = \mathbb{C}1$. He will no doubt explain them in his talk. Pimsner and Popa have shown that for property T II_1 factors the set of subfactors up to inner conjugacy is countable! For factors like the group von Neumann algebras of free groups the existence problem for index values of subfactors, even < 4, is wide open.

In this talk I shall largely avoid internal questions of the theory and talk about relations with other parts of mathematics and physics. The exception will be the first topic which I feel to be of fundamental importance.

I Commuting squares

A <u>commuting square</u> will be a quadruple $\begin{matrix} B_0 \subseteq B_1 \\ \cup \quad \cup \\ A_0 \subseteq A_1 \end{matrix}$ of finite von Neumann algebras together with a faithful normal trace tr on B_1 with the (equivalent) properties.

a) $E_{B_0}(A_1) = A_0$ (E_{B_0} is the trace preserving conditional expectation from B_1 to B_0).

b) $L^2(B_0,tr)$ and $L^2(A_1,tr)$ are commuting subspaces of the Hilbert space $L^2(B_1,tr)$ with intersection $L^2(A_0,tr)$.

c) $E_{A_0}(x\,y) = E_{A_0}(x)E_{A_0}(y)$ if $x \in B_0$ and $y \in A_1$. (I owe this last condition to de la Harpe).

It is an easy exercise to see that if A,M and N are finite von Neumann algebras with traces, then

$$\begin{matrix} A \otimes M \otimes 1 & \subseteq & A \otimes M \otimes N \\ \cup & & \cup \\ A \otimes 1 \otimes 1 & \subseteq & A \otimes 1 \otimes N \end{matrix}$$

is a commuting square. Popa's notion of orthogonal pairs of algebras is exactly that of commuting squares with $A_0 = \mathbb{C}\,1$. The notion of commuting squares has evolved along with subfactors. It appears explicitly in Wenzl's thesis and in Pimsner and Popa's "Entropy and index for subfactors" paper.

One reason for the importance of this notion for subfactors is the following. If $\begin{matrix} B_0 \subseteq B_1 \\ \cup \quad \cup \\ A_0 \subseteq A_1 \end{matrix}$ is a commuting square, then consider the basic construction for the pair $B_0 \subseteq B_1$. One has von Neumann algebras

$$\begin{matrix} B_0 \subseteq B_1 \subseteq B_2 = \langle B_1, e_{B_0} \rangle \\ \cup \quad \cup \quad \cup \\ A_0 \subseteq A_1 \subseteq A_2 = \{A_2, e_{B_0}\}'' \end{matrix}$$

The existence of a faithful normal trace on B_2 extending tr which makes

$$\begin{matrix} B_1 \subseteq B_2 \\ \cup \quad \cup \\ A_1 \subseteq A_2 \end{matrix}$$

is a commuting square is, under suitable irreducibility assumptions, the same as the <u>Markov property</u> for that trace, i.e. $tr(xe_{B_1}) = \tau\, tr(x)$ for

$x \in B_1$ and some $\tau \in \mathbb{R}^+$. It is well established that the Markov condition will continue if one iterates the construction so we have, automatically from the first square a whole tower

$$\begin{array}{ccccccc} B_0 \subseteq & B_1 \subseteq & B_2 \subseteq & .. & B_i \subseteq \\ \cup & \cup & \cup & & \cup \\ A_0 \subseteq & A_1 \subseteq & A_2 \subseteq & .. & A_i \subseteq \end{array}$$

of commuting squares and a privileged faithful normal trace on B_i for all i. Under the same irreducibility assumptions the unions UB_i and UA_i give II_1 factors $N \subseteq M$ when the GNS construction is applied.

Moreover $\begin{array}{cc} B_i \subseteq & M \\ \cup & \cup \\ A_i \subseteq & N \end{array}$ is a commuting square for all i which allows one to approximate the basic construction for $N \subseteq M$ by those for $A_i \subseteq B_i$, and in particular to calculate $[M:N]$. It would be important to identify $N' \cap M$ in this process. Wenzl has made important contributions to this.

A suggestive reason for the interest of commuting squares comes from the connection between subfactors and braids. For some while it has been known that, if one considers the projections e_i in the tower for a subfactor $N \subseteq M$ where $M_{i+1} = \langle M_i, e_{i-1}\rangle$, $M_1 = N$, $M_2 = M$, then defining $g_i = te_i - (1-e_i)$, $2 + t + t^{-1} = [M:N]$, yields representations of the union of the braid groups B_∞ $(= \langle \sigma_1, \sigma_2, \ldots \mid \sigma_i\sigma_{i+1}\sigma_i = \sigma_{i+1}\sigma_i\sigma_{i+1}$, $\sigma_i\sigma_j = \sigma_j\sigma_i$ if $|i-j| \geq 2\rangle)$ by sending σ_i to g_i. But the change of variables $e_i \to g_i$ seems <u>completely arbitrary</u>. However if $[M:N] \leq 4$, g_i is <u>unitary</u> and one may show that $u = g_1^{\pm 1}$ are the unique unitaries in $\text{alg}(1,e_N)$ for which the square

$$\begin{array}{ccc} u M u^{-1} & \subseteq & \langle M,e_N\rangle \\ \cup & & \cup \\ N & \subseteq & M \end{array}$$

is commutative. This suggests considering in general the collection of subfactors P between N and $\langle M,e_N\rangle$ for which

$$\begin{array}{cc} P \subseteq & \langle M,e_N\rangle \\ \cup & \cup \\ N \subseteq & M \end{array}$$

is a commuting square, as an invariant for $N \subseteq M$. In concrete examples it should not be too hard to calculate, and is probably in general related to the braid group.

I believe that commuting squares should be studied in their own right and have their own rich structure.

II Vertex models

Given a graph G with edges $E = \{e\}$ and vertices $S = \{s\}$, one defines vertex models on G by choosing a natural number n_e for each edge. A state of the system is then a function $\sigma : E \to \mathbb{N}$ with $\sigma(e) \le n_E$. Each vertex is then surrounded by a certain configuration of numbers on the edges incident to it. One postulates an energy $E_\sigma(s)$ which depends only on this configuration.

The model is then completely defined and one defines the partition function Z to be

$$Z_G = \sum_{\sigma} \prod_{s \in S} \exp(-\beta E_\sigma(s)) \quad .$$

The above discussion was very abstract. In physical situations the graph is typically a lattice in $\mathbb{R}^n$. We shall be most interested for the moment in the case of $\mathbb{Z} \oplus \mathbb{Z}$ in $\mathbb{R}^2$, but note first that the sum will not converge so one is interested in approximating the lattice by finite pieces and considering the asymptotic behaviour of Z as the pieces tend to the whole system. To do this the technique of transfer matrices has been most successful. It is very easily understood in one dimension. Here the lattice is $\mathbb{Z}$ pictured as —o—o—o—o—o—...... Fix $n_e = n$ and the partition function becomes

$$Z_k = \sum_{\sigma} \prod_{i=-k}^{k} w(a,b) \quad \text{where} \quad w(a,b) = \exp(-\beta E_i(a,b)) \ ,$$

a and b being the numbers assigned by σ to the edges joining i to i-1 and i+1 respectively. (E_i is assumed independent of i).

In general the approximation of the whole lattice by finite pieces will create some difficulties on the boundary. The easiest way to handle them in this case is to assume that the states are periodic. Then Z_k becomes

$$\sum_{i_{-k}, i_{-k+1}, \ldots, i_{k-1}} \prod_{p=-k}^{k-1} w(i_p, i_{p+1}) \quad \text{where} \quad i_k = i_{-k} \ .$$

One immediately sees $Z_k = \text{trace}(T^{2k-1})$ where T is the $n \times n$ matrix whose (a,b) entry is $w(a,b)$. The problem is now completely solved by diagonalizing T, as the largest eigenvalue determined the asymptotic behaviour.

Let us now turn to the two dimensional case where the lattice/ graph is

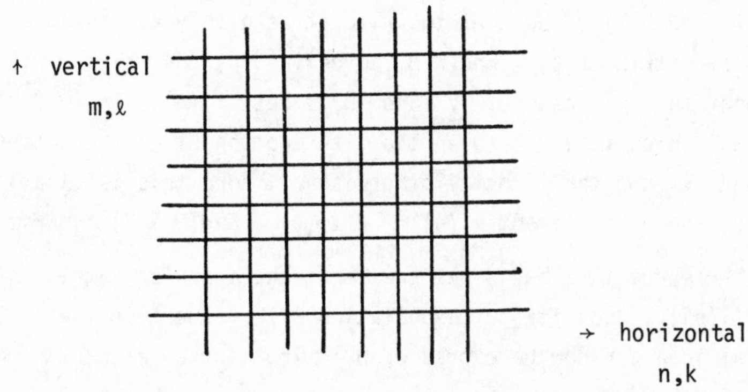

The idea is to treat this as a one dimensional system in which the rows become the atoms (vertices) of the lattice and to approximate by increasing rectangles. The complication of course is that the number of possible states for the new atoms (and hence the size of the transfer matrix) grows exponentially with the horizontal dimension of the approximating rectangle. Diagonalizing the transfer matrix may then be extremely difficult.

In order to diagonalize the transfer matrix it is helpful to have as many natural operators that commute with it as possible. One possible source of such operators (emphasized by Baxter) is the transfer matrix at other values of the parameter θ (what parameter? β , maybe others). As in the one dimensional case we will suppose the weights w $(= \exp(-\beta E))$ are the same at every vertex. Note that now they will depend on four indices, $w(a,b \mid x,y)$ corresponding to the following vertex configuration:

a | y b
——+——
  | x

The question we ask is the following: is there a (local) condition on the $w(a,b \mid x,y)(\theta)$ which guarantees that the row to row transfer matrices commute for different values of θ ? This will be seen to lead to a simple commuting square and the famous Yang-Baxter equation (YBE). It is convenient to assymetrize the situation by choosing a horizontal vector space V of dimension n and a vertical one W of dimension m (in the abstract graph terminology, $n_e = n$ if e is horizontal, $n_e = m$

if e is vertical). We will approximate by a $k \times \ell$ rectangle and form the vector space $H = \otimes^k W$. The transfer matrix $T(\theta)$ will then be an element of $\mathrm{End}(H,H)$. We want to write it as nicely as possible in terms of the local matrix $t(\theta) \in \mathrm{End}(W \otimes V)$ defined by the weights $w(a,b|x,y)$ with respect to some bases of V and W. Let $t_i(\theta) \in \mathrm{End}(H \otimes V)$ be the operator which is $t(\theta)$ between the ith component of the tensor product and V and the identity otherwise (I hope this is clear). Then one may consider the element $\tilde{T}(\theta) = \prod_i t_i(\theta) \in \mathrm{End}(H \otimes V)$. Writing out $\tilde{T}(\theta)$ with respect to a basis one sees that the transfer matrix $T(\theta)$ is $E_{\mathrm{End}(H)\otimes 1}(\tilde{T}(\theta))$, conditional expectation with respect to the trace. The fundamental idea can now be expressed by: look for an element $R(\theta,\theta') \in \mathrm{End}(V \otimes V)$ such that

$$(*) \qquad R(\theta,\theta')t_{12}(\theta)\, t_{13}(\theta')\, R(\theta,\theta')^{-1} = t_{13}(\theta')t_{12}(\theta)$$

where t_{12} on $\overbrace{W \otimes V} \otimes V$ acts as indicated and t_{13} acts as $\overbrace{W \otimes V \otimes V}$. One then sees that property (c) for commuting squares with $A = \mathrm{End}(H)$, $M = \mathrm{End}(V)$ $N = \mathrm{End}(V)$ in the first example of commuting squares guarantees that $T(\theta)T(\theta') = T(\theta')T(\theta)$. Such vertex models are said to be exactly solvable.

The similarity with the subfactor picture is quite remarkable since $\mathrm{End}(H)$ naturally tends to a II_1 factor (or a Powers factor) and although the subfactor obtained by tensoring with $\mathrm{End}\, V$ is not too exciting it may become more so if a group is acting. Also, vertex models are just one kind of model, and one is to expect more elaborate subfactors from other models (Potts, Andrews-Baxter-Forrester, Pasquier).

In fact if $\dim V = 2$ the analogy with ($[M:N] = 4$) subfactors is perfect. If $M = N \otimes M_2(\mathbb{C})$ then $\langle M,e_N\rangle$ can be naturally identified with $N \otimes M_2(\mathbb{C}) \otimes M_2(\mathbb{C})$ and $t = 1$ so that $g = 2e_N - 1$ is actually the permutation on $V \otimes V$, $v \otimes v' \to v' \otimes v$ (up to sign) so that $g\, M g^{-1} = N \otimes 1 \otimes M_2(\mathbb{C})$. Thus the commuting square coming from the braid group is the same as the one used in statistical mechanics. If $\dim V > 2$ the permutation matrix will again do the job so this lends even more weight to replacing the braid group structure by commuting squares in general.

In fact it seems likely that subfactors give, more or less canonically, statistical mechanical models of an exactly solvable kind. One should look at the relative commutants in the tower $N \subseteq M \subseteq M_1 \subseteq M_2 \subseteq \dots$. Such models have been worked out in some cases by Pasquier. They generalize the Andrews-Baxter-Forrester models. One begins with a

Coxeter graph and decrees that the states of the atoms are the vertices of the Coxeter graph. The model then only allows states with the property that neighbouring atoms are assigned neighbouring vertices on the Dynkin diagram. The weights $w(a,b \mid x,y)$ (a,b,x,y are the vertices of neighbouring atoms - this is not a vertex model) are related to eigenvectors of the incidence matrix of the Coxeter graph - a condition forced by integrability).

The same structure occurs in Ocneanu's analysis of subfactors.

III The Yang Baxter equation

In the preceding discussion of vertex models there is a consistency condition.

Considering different ways to reorder $V \otimes V \otimes V$ one is led to the following likely condition on R :

(YBE) $$R_{12}(\theta,\theta')R_{13}(\theta',\theta'')R_{23}(\theta'',\theta) = R_{23}(\theta'',\theta)R_{13}(\theta',\theta'')R_{12}(\theta,\theta') .$$

Moreover the direction can be reversed: given a solution of YBE depending only on $\theta-\theta'$ one can construct $t(\theta)$ and (YBE) implies (*).

Thus one is led to look for solutions of the YBE. Note the similarity with the braid group relations. The simplest solution of real interest is that corresponding to the "ice-type" or six vertex model solved by Lieb. One has $\dim V = 2$. The R matrix is then a 4×4 matrix, in fact up to a scalar the following

$$\begin{pmatrix} 1 & 0 & 0 & 0 \\ 0 & b & c & 0 \\ 0 & c & b & 0 \\ 0 & 0 & 0 & 1 \end{pmatrix}, \quad \begin{array}{l} b = \sinh(\theta)/\sinh(\lambda-\theta) \\ c = \sinh(\lambda)/\sinh(\lambda-\theta) \\ \quad (\theta \text{ as below}) \end{array}$$

In field theory the YBE has a simple geometric interpretation as factorizability of the S-matrix. The idea is that an n-particle interaction can be decomposed as a composition of 2-particle interactions. Then if there are 3 particles 1,2,3 and $S_{ij}(\theta)$ is the scattering operator for an interaction between i and j with relative "velocity" θ , there are 2 ways to decompose the 3 particle interaction according to the picture. One reads off the YBE

$$S_{12}(\theta)S_{13}(\theta'+\theta)S_{23}(\theta') = S_{23}(\theta')S_{13}(\theta'+\theta)S_{12}(\theta)$$

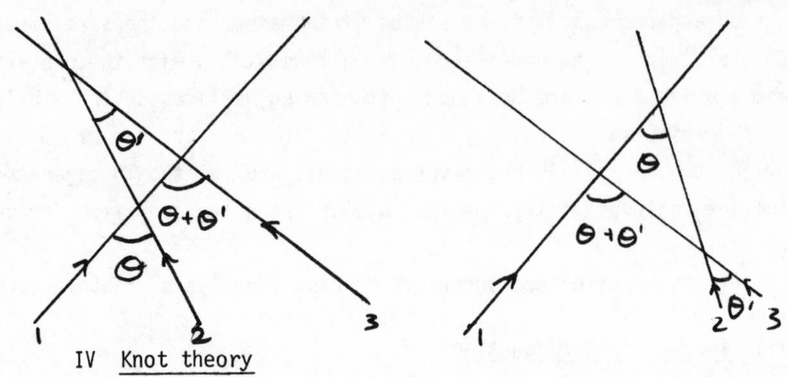

IV Knot theory

The easiest way to visualize knots and links in three dimensions is by way of (generic) plane projections. If one records crossing data one gets a more than adequate picture of a link, e.g.

Since there are only double points as singularities of the projection, the situation locally resembles that of a $\mathbb{Z} \oplus \mathbb{Z}$ lattice so one may consider vertex models with no θ-dependence on the projection and ask the question: is it possible that the partition function is an invariant, i.e. depends only on the link up to isotopy in 3 space? (The assymetrizing of the transfer matrix approach shows how to deal with the different components of a link differently - just choose a different vector space for each component.)

Such an invariant would be of no interest since we have not taken into account the crossing data and all that could remain would be the number of components. However if the link is oriented, crossings can be given a sign + = , - = , and it is natural to choose a different set of weights w at crossings, w_+ for positive, w_- for negative. Then it is possible that non-trivial invariants could be obtained from partition functions. Without some background this sounds like absurdly wishful thinking but right from the braid - von Neumann algebra discovery of new knot polynomials it has been known that there is a connection with statistical mechanics and Kauffman actually defined V_L using a model on knot projections (not obviously a vertex model). In fact it is entirely conceivable that the collection of invariants obtained in the simple minded way outlined above is faithful on oriented links! There are certainly enough to determine V. The first was obtained by Lipson and corresponds to the following R-matrix (we assume that t is given by R from now on)

$$\begin{pmatrix} \cosh\lambda & 0 & 0 & \sinh\lambda \\ 0 & -\sinh\lambda & -\cosh\lambda & 0 \\ 0 & -\cosh\lambda & -\sinh\lambda & 0 \\ \sinh\lambda & 0 & 0 & \cosh\lambda \end{pmatrix}$$

As shown by Lickorish and Lipson, the link invariant so obtained is the moment generating function for the random variable on the sublinks of the link given by linking number with the rest of the link.

Further invariants obtained by Lipson are specializations of V_L corresponding to the element of $\mathrm{End}(V \otimes V)$ given by a Euclidean inner product on V.

But to return to the story, one must find conditions on the two solutions w for the partition function Z_L to be an invariant. Three dimensional isotopy is translated into link diagrams by the Reidemeister moves, which come in three types according to the number of crossings involved.

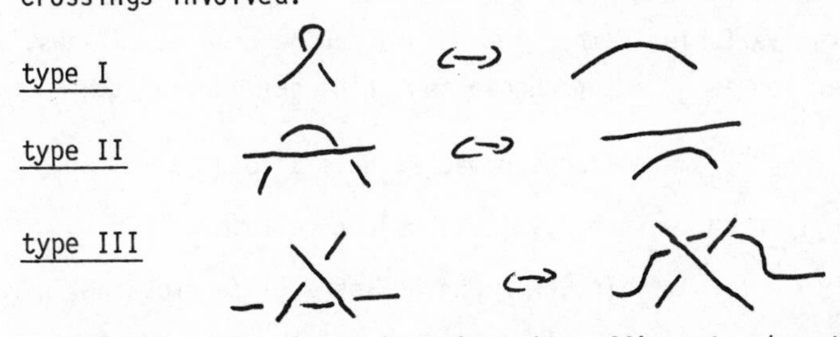

To show that Z_L is an invariant it suffices to show that it does not change under moves I, II, III. Let us examine what this implies for the R matrix.

Type I By some ideas emphasized by Kauffman, one may ignore type I.

Type II In fact there are two situations to consider

and (and their mirror images)

(A) (B)

The thing is that the diagrams for  and

have less states than those for (A) and (B). So the condition must be that a sum over internal variables cancels unless

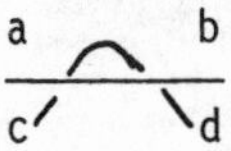

a b

c d , $a = b$, $c = d$. This is fairly obviously just a way of

saying that the W_+ R-matrix, call it R, is the inverse of the w_- R-matrix provided one identifies R as an endomorphism of $V \otimes V$ in the appropriate manner. If this is done for II_A one finds that the condition coming from II_B is different. In fact if we write $End(V \otimes V) = End(V) \otimes End(V)$ there are transpose operations $t_1 =$ (transpose) $\otimes$ id and $t_2 =$ id $\otimes$ (transpose) and the condition for invariance is $R^{t_1}(R^{-1})^{t_2} = 1$. I have not yet made much sense of this condition but it must be natural.

Type III All the various possible orientations etc., can be reduced to precisely the YBE (with no θ-dependence).

So we get the result that to every invertible matrix $R \in End(V \otimes V)$ satisfying $R^{t_1} (R^{-1})^{t_2} = 1$ and the YBE there is an oriented link invariant.

If one limits oneself to the theory above it is rather difficult even to obtain V_L but, taking a cue from the physics, one can try to use the θ variable in the YBE. This can be done as follows. Make sure your projection is C^∞ and choose one point per edge of the projection ( has 6 edges) . Then suppose we have a solution $R_\pm(\theta_1,\theta_2)$ of YBE given by $w_\pm(a,b|x,y)(\theta_1,\theta_2,\theta)$ and define

$$Z_L = \sum_{\text{states}} \prod_{\text{crossings}} w\,(a,b|x,y)(\theta_1,\theta_2,\theta)$$ where for a given crossing the values θ_1 and θ_2 are the changes in angle between the points on the left incoming and right outgoing, and right ingoing and left outgoing curves respectively. And θ is the change in angle between the left outgoing and right outgoing curves.

Once again we can write down the conditions for this Z_L to be a link invariant. One now has to consider dependence on the choice of the points. This turns out to be a simple property involving $R(\theta_1+\alpha,\theta_2,\theta)$. Once Z_L is independent of the points it is an invariant of planar isotopy and one may consider invariance under Reidemeister moves as before. Without getting into detail let me say simply that V_L is obtained from a solution corresponding precisely to the R matrix for the ice-type model.

It is also intriguing that one can see the relevance of t-matrices in this picture as it is possible to find choices (R,t) for which one may choose as many points as one likes on the knot and use the t-matrix inbetween points that do not have a crossing and the R-matrix between points that do. The equation $Rt_1(\theta)t_2(\theta')R^{-1} = t_2(\theta')t_1(\theta)$ is then implied by the second Reidemeister move

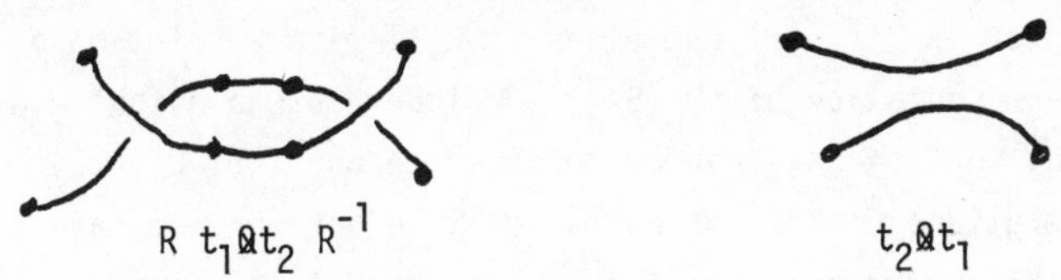

V <u>Quantum groups</u>

Thus far we have begged the important question of how one is to come up with solutions of YBE. The payoff is good, as we have seen, to every solution we are likely to get an exactly solvable vertex model a sub-factor, a knot invariant and a factorizable S-matrix. (In fact there will be a completely integrable $1 + 1$ dimensional quantum field theory as well.) We have hinted at the existence of interesting solutions. Quantum groups are a machine which was developed in order to construct a wide variety of solutions of YBE. The simplest case came from a study of the following commutation relations (all this theory was pioneered by Fadeev and the Russian school, culminating in Drinfeld's formalism of quantum groups. Woronwicz was led to simlar objects for different reasons):

$$s\ell(2,h): \quad \begin{aligned}[H,E] &= 2E \\ [H,F] &= -2F \\ [E,F] &= (\sinh(hH/2))/(h/2)\end{aligned}$$

If $h \to 0$ we recover $s\ell_2$ in its usual presentation. For non-zero h we must work on the level of the enveloping algebra. The point is that these relations admit a tensor product operation: given H,E,F on V and W satisfying $s\ell(2,h)$, one may define H,E,F on $V \otimes W$ by

$$\begin{aligned}H &= H \otimes 1 + 1 \otimes H \\ E &= E \otimes \exp(\tfrac{hH}{4}) + \exp(\tfrac{-hH}{4}) \otimes E \\ F &= F \otimes \quad \text{"} \quad + \quad \text{"} \quad \otimes F\end{aligned}$$

then the new H,E,F satisfy $s\ell(2,h)$. Changing the sign in the exponential gives another solution H',E',F' and the R-matrix comes in as an element of $\mathrm{End}(V \otimes W)$ conjugating H,E,F to H',E',F' . More formally one defines the associative algebra $U_h(S\ell_2)$ as the (suitably completed to take care of the power series) algebra with presentation $s\ell(2,h)$. The tensor product operation is then the existence of a co-product $\Delta : U_h(s\ell_2) \to U_h(s\ell_2) \otimes U_h(s\ell_2)$ and the universal R matrix will be an element of $U_h(s\ell_2) \otimes U_h(s\ell_2)$ satisfying $R\Delta(x)R^{-1} = \sigma o \Delta(x)$ where $\sigma \in \mathrm{Aut}(U_h(s\ell_2) \otimes U_h(s\ell_2))$ is defined by $\sigma(x\otimes y) = y\otimes x$. In fact an exact formula for R is very useful as it will give a solution of YBE for every representation of $U_h(S\ell_2)$. Jimbo has shown that every finite dimensional rep. of $s\ell_2$ can be deformed to one for $U_h(s\ell_2)$ so we will get solutions of YBE for every positive integer. I have checked that the YBE solutions are of the right kind to give knot invariants so one has polynomials for every irrep of $s\ell_2$. They will be determined by V of cables of a knot but this should not be construed as a negative fact since the object is to understand the polynomials and not to create more.

Turaev has worked out the explicit formulae for the so(2n+1) , sp(2n) so(2n) series in their identity representations. One obtains specializations of the Kauffman polynomial. $s\ell_n$ in general gives specializations of the 2-variable Alexander polynomial P_L .

For the record, Drinfeld gives the following formula for the R-matrix in $U_h(s\ell_2) \otimes U_h(s\ell_2)$:

$$R = \sum_{k=0}^{\infty} e^{-kh/2} \prod_{r=1}^{k} \frac{(e^h-1)}{(e^{rh}-1)} \{\exp \tfrac{h}{4}[H\otimes H + k(H\otimes 1 - 1\otimes H)]\} E^k \otimes F^k h^k .$$

For subfactors, the interest in quantum groups is quite direct. Given $R \in \mathrm{End}(V \otimes V)$ one considers the subalgebra of $\otimes^\infty \mathrm{End}(V)$ generated by $R_1,R_2,R_3 \ldots.$ which is contained in the commutant of the quantum group action. Then $R_2,R_3 \ldots$ will generate a subalgebra.

VI Field theory

One of the most important aspects of statistical mechanical models is that there should exist a non trivial quantum field theory as the lattice spacing tends to zero. If it exists it can be defined in terms of the n-point or correlation functions which will be defined for a finite lattice spacing as

$$\langle f(Z_1) \ldots f(Z_n)\rangle = \frac{\sum_\sigma f(Z_1)\ldots f(Z_n)\exp(-\beta E(\sigma)}{Z}$$

where f is the variable of interest and $Z_1 \ldots Z_n$ are lattice points. If the lattice spacing tends to zero keeping $Z_1,..,Z_n$ fixed in $\mathbb{R}^2$ there is hopefully a sense in which the limit of $\langle f(Z_1)..f(Z_n)\rangle$ exists. Provided these limit functions satisfy certain conditions (Osterwalder Schrader axioms) one is then guaranteed the existence of a field theory (operator valued distributions on a Hilbert space etc.) for which the $\langle f(Z_1).... f(Z_n)\rangle$ are the boundary values of an analytically continued n point function, $\langle\Omega|\phi(x_1,t_1) \ldots \phi(x_n,t_n)|\Omega\rangle$. Thus the Euclidean field theory is that naturally comes out of statistical mechanics. The most interesting points at which to let the lattice spacing go to zero are the critical points for which long range phenomena occur. The homogeneity of the system suggests that the field theory will have some kind of invariance under local conformal transformations.

This possibility is even more strongly suggested by the knot theoretic approach where the scaling limit could be approached is a totally haphazard fashion

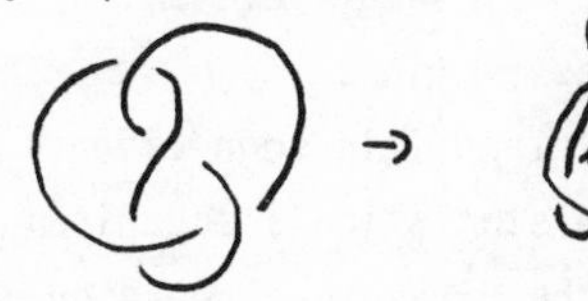

which would entirely forget any privileged directions.

In any case a local conformal transformation (away from zero, ∞) will have the form $z \to \sum_{-\infty}^{+\infty} a_n z^n$ and one expects the Lie algebra of these symmetries to act on the field theory. Since the action may be projective, one obtains an action of the Virasoro algebra which has presentation on generators c, $\{L_n, n \in \mathbb{Z}\}$

$$[L_n, L_m] = (n-m)L_{n+m} + c\,\frac{\delta_{n,-m}(n^3-n)}{12}$$

$$[C, L_m] = 0 \quad \forall\, m \quad .$$

In an explicit theory theory one may calculate L_n from the so called "stress energy tensor" $T(z) = \sum_{n\in\mathbb{Z}} L_n z^n$. The element c is the so-called central charge and in an irreducible representation it will be a scalar. If the theory is physical the L_n's will be subject to some unitary condition on a Hilbert space. It appears to be $L_n^* = L_{-n}$, and Friedan, Qiu and Schenker have shown that this imposes restrictions on the scalar c in an irreducible representation. Namely $c > 0$ and if $c < 1$ then $c = 1 - \frac{6}{m(m+1)}$ for some integer $m \geq 3$. There is a general feeling that these discrete values of c correspond in some way to the $4\cos^2\pi/n$ values for the index for subfactors. The feeling is reinforced by many observations, perhaps the most convincing of which is that the algebra of e_i's for $[M:N] = 4\cos^2 \pi/n$ does occur as transfer matrices in statistical mechanical systems for which I am told, the corresponding central charge is $1 - \frac{6}{n(n-1)}$. It is also true that the original proof of restrictions on index values is morally the same as Friedan Qiu and Schenker's proof - they consider a sesquilinear form on finite dimensional spaces of increasing dimension defined by the Hilbert space inner product restricted to the span of $L_{-1}, L_{-2}, \ldots L_{-k}$ applied to an eigenvector of L_0. The constraint of positivity of the inner product gives the restrictions. In the subfactor proof one considers the inner product given by the trace on the sequence of vector spaces of increasing dimension given by $\mathrm{alg}(1, e_1, \ldots e_k)$. Positivity of this inner product gives the restrictions.

It seems to be important to understand this connection. Since the field theory is in fact determined by the statistical mechanical model, one is naturally led to consider actions of the Virasoro algebra on the algebra generated by the e_i's. The "right" action would be the one which gave the right action on the field theory in the scaling limit though at this stage one would settle for the right central charge.

An important first step has been taken by Connes and Evans who used as a guiding principle the fact that for $[M:N] = 2$, the e_i algebra is actually the even CAR or Clifford algebra where explicit formulae are known for the field theory. By solving certain equations in e_i's they were able to generalize these formulae to obtain actions of Virasoro (and more) for arbitrary $[M:N]$. But it turns out that the central charge is always the same as for $[M:N] = 2$.

In a useful paper for mathematicians, Tsuchiya and Kanie concentrate on a specific model (Σ-model on $SU(2)$ with Wess-Zumino-Witten term) treated by Knizhnik and Zamolodchikov and express its mathematical content in purely mathematical terms. (Representations of the affine $s\ell_2$

algebra and Virasoro algebra). They explicitly calculate the n point functions $\langle\phi(Z_1)\ ..\ \phi(Z_n)\rangle$ and show that they satisfy a differential equation with regular singular points on $\mathbb{C}^n$, the singularities being on $\Delta = \{(Z_1..Z_n) \mid Z_i = Z_j \text{ for some } i \neq j\}$. Then they calculate the monodromy representation of $\pi_1(\mathbb{C}^n - \Delta)$ and find precisely the braid group representations that come from $4\cos^2 \pi/k$ subfactors. But the central charge is of the form $\frac{3k}{k+2}$. Unfortunately there do not seem to be any field theory models realizing all the $1 - \frac{6}{m(m+1)}$ values without going through the scaling limit.

It is important to note, in fact, that even the existence of the $1 - \frac{6}{m(m+1)}$ values of c requires a non-trivial trick due to Goddard and Olive: affine Lie algebras $\mathcal{Y}$ are acted on by the Virasoro algebra and the Virasoro operators can be written down explicitly in the enveloping algebra of the affine Lie algebra (the "Segal Sugawara" form)

$$\left(\text{If } (T^a_m, T^b_n) = \sum_c C^{ab}_c T^c_{m+n} + k\, m\, \delta_{m,-n}\, \delta_{a,b} \quad \text{then if}\right.$$

$$L_n = \text{const} \sum_{n \in \mathbb{Z}} :T^a_m T^a_{n-m}: \left(\begin{array}{ll} :T^a_m T^a_n: & = T^a_m T^a_n \quad m < 0 \\ & = T^a_n T^a_m \quad m > 0 \end{array} \right),$$

L_n satisfies Virasoro with some c). Given the appropriate subalgebra h of $\mathcal{Y}$, simply subtracting the Virasoro operators for $\hat{h}$ from those for $\hat{\mathcal{Y}}$ gives a new representation in which the central charge is the difference of the central charges. No such operation is known on subfactors which will "subtract" indices but note that the first proof of the existence of subfactors of index $4\cos^2 \pi/n$ used a pair of finite dimensional semi-simple <u>associative</u> algebras. The analogy remains.

Finally we would like to draw the audience's attention to the "operator algebras" of quantum field theory. Associated with Gell-Mann one has the equal time commutation relations

$$[A(x_0,x),\ B(x_0,y)] = \sum_n D_n(x-y)\ O_n(x_0,y)$$

where the $O_n(x)$ are local fields and D_n are δ functions and their derivatives. The "algebra" is defined by the particular choices of D_n and O_n. Wilson proposed a generalization where one supposes in fact that

$$A(x)B(y) = \Sigma\, C_n(x-y)\; O_n(x)$$

where x is a space-time coordinate and $C_n(x-y)$ are only powers of $x-y$. Thus the $C_n(x-y)$ are structure constants for an "algebra" and one idea was to use associativity constraints to help in the calculation of the n-point functions. Polyakov applied this in the case of conformal field theory.

It seems important to try to understand the many examples of "operator algebras" in terms of our operator algebras = C^* and von Neumann algebras.

The following references may be useful in understanding the text.

von Neumann algebras

V. Jones. "Index for subfactors". Invent. Math 72 (1983) 1-25.

H. Wenzl. "Representations of Hecke algebras and subfactors". Thesis, University of Pennsylvania (1985).

M. Pimsner and S. Popa. "Entropy and Index for subfactors". Ann. Sci. Ec. Norm. Sup. 19 (1986) 57-106.

Knot theory

D. Rolfsen. "Knots and Links". Publish or Perish, Berkeley, 1976.

L. Kauffman. "State models and the Jones polynomial" to appear in topology.

V. Jones. "Hecke algebra representations of braid groups and link polynomials". To appear, Annals of Math.

Physics

J. Lepowsky, S. Mandelstam and I.M. Singer. "Vertex operators in Mathematics and Physics". MSRI publications, Springer Verlag (1984).

R. Baxter. "Exactly solved models in Statistical mechanics". Academic Press, London 1982.

A. Tsuchiya and Y. Kanie. "Vertex operators on conformal field theory on $\mathbb{P}^1$ and monodromy representations of braid groups". To appear in "Conformal field theory and solvable Lattice models", Advanced Studies in pure mathematics, Konokuniya.

Other

V. Drinfeld. "Quantum groups". To appear, proc. ICM 86.

QUANTIZED GROUPS, STRING ALGEBRAS AND GALOIS THEORY FOR ALGEBRAS

Adrian Ocneanu
Department of Mathematics, The Pennsylvania State University,
215 McAllister Building, University Park, Penn. 16802, U.S.A.

ABSTRACT

We introduce a Galois type invariant for the position of a subalgebra inside an algebra, called a paragroup, which has a group-like structure. Paragroups are the natural quantization of (finite) groups. The quantum groups of Drinfeld and Fadeev, as well as quotients of a group by a non-normal subgroup which appear in gauge theory have a paragroup structure.

In paragroups the underlying set of a group is replaced by a graph, the group elements are substituted by strings on the graph and a geometrical connection stands for the composition law. The harmonic analysis is similar to the computation of the partition function in the Andrews-Baxter-Forrester models in quantum statistical mechanics (e.g. harmonic analysis for the paragroup corresponding to the group $\mathbb{Z}_2$ is done in the Ising model.) The analogue of the Pontryagin van Kampen duality for abelian groups holds in this context, and we can alternatively use as invariant the coupling system, which is similar to the duality coupling between an abelian group and its dual.

We show that for subfactors of finite Jones index, finite depth and scalar centralizer of the Murray-von Neumann factor R the coupling system (or alternatively the paragroup) is a complete conjugacy invariant. In index less than 4 these conditions are always satisfied and the conjugacy classes of subfactors are rigid: there is one for each Coxeter - Dynkin diagram A_n and D_{2n}, and there are two anticonjugate but nonconjugate for each of the E_6 and E_8 diagrams. Thus underlying the rigidity of the Jones index there is a crystal-like rigidity of the position of subfactors of R.

INTRODUCTION

THE PROBLEM

A subalgebra A_1 of an algebra B_1 and a subalgebra A_2 of an algebra B_2 are conjugate (or have the same position) if there is an isomorphism $\theta: B_1 \rightarrow B_2$ with $\theta(A_1) = A_2$. In what follows we study the possible positions of a subalgebra into an algebra, i.e. the conjugacy classification of subalgebras. This problem is an analogue of the classical Galois theory for fields, although in our case we want to impose no a priori restrictions similar to the normality and separability of field extensions. It is convenient to study the problem in its technically simplest form; we shall consider a subalgebra A of a complex algebra B which has scalar centralizer in B and finite index in B, i.e. B as a module over A is finitely generated. Since we are interested in the position of the subalgebra, the isomorphism class of the subalgebra in itself should be ideally irrelevant; moreover the algebra should be rich enough to have subalgebras in interesting positions.

The algebra used in what follows is the Murray-von Neumann factor R, or the hyperfinite II_1 factor in von Neumann's classification, also called in the sequel the elementary von Neumann algebra. The algebra R, which is the weak closure of the Clifford algebra of the real separable Hilbert space, is a factor (i.e. it has scalar center) which has very many symmetries (e.g. any locally compact group can act outerly on R.) A strong uniqueness theorem of Connes implies that any closed subalgebra of R which is a factor, i.e. has trivial center, is isomorphic either to $\mathrm{Mat}_n\mathbb{C}$ or to R itself. Thus any finite index subfactor N of R is isomorphic to R, and all the information in the inclusion $N \subset R$ comes from the relative position of N in R and not from the structure of N. The algebra R has a canonical trace tr with the trace of idempotents varying continuously between 0 and 1 (R is a continuous geometry, or a type II_1 factor in the sense of von Neumann,) and is the weak closure of an ascending union of finite dimensional subalgebras (i.e. R is hyperfinite, or approximately finite dimensional.) The classification theorem of Murray and von Neumann states that any hyperfinite II_1 factor is isomorphic to R; in our context this guarantees that the closure of all finite dimensional constructions done below will bring us back to R.

If $A \subset B$ with B elementary, then B is a projective left module over A with index $[B : A] = \dim {}_A B \in [1, +\infty]$ computed by means of the trace. The index was introduced (with a different but equivalent definition) by V. Jones, who proved a remarkable rigidity theorem: if $[B : A] < 4$, then $[B : A] \in \{4 \cos^2(\pi/n);\ n = 3,4,\ldots\}$. We shall show in the sequel that underlying the rigidity of the index there is rigidity for the conjugacy class of the subalgebra A: for any $n \geqq 3$ there are at most 4 conjugacy classes of subfactors of R with index $4 \cos^2(\pi/n)$. The construction will also explain and give a short proof of the rigidity of the Jones index.

THE GALOIS FUNCTOR

We introduce a conjugacy invariant for a finite index subfactor N of a factor M, consisting of a pair of graphs, an involutive bijection of their vertices, and a connection, which is a cohomological invariant defined on small loops in the graphs called cells.

Consider the algebraic tensor products $M^{\otimes n} = M \otimes_N M \otimes_N \cdots \otimes_N M$ (n times) where $M^{\otimes 0} = N$. Let $\mathfrak{G}$ be a bipartite graph having even vertices $\mathfrak{G}^{(0)}_{\text{even}}$ the equivalence classes of irreducible $N - N$ subbimodules of $M^{\otimes n}$, $n = 0,1,\ldots$ and odd vertices $\mathfrak{G}^{(0)}_{\text{odd}}$ the equivalence classes of irreducible $M - N$ subbimodules of $M^{\otimes n}$, $n = 1,2,\ldots$ with k edges between the classes o ${}_N X_N$ and ${}_M Y_N$ if the restriction ${}_N Y_N$ contains ${}_N X_N$ k times. The class o the identity bimodule ${}_N N_N$ is distinguished and marked $*_{\mathfrak{G}}$. The graph $\mathfrak{G}$ i.

connected, and the finiteness of the index guarantees that the graph $\mathcal{G}$ is locally finite. The graph $\mathcal{G}$ is called the induction - restriction graph of $N \subset M$. Construct analogously the restriction - induction graph $\mathcal{H}$ of $N \subset M$, with even and respectively odd vertices the equivalence classes of irreducible $M - M$ and respectively $N - M$ subbimodules of $M^{\otimes n}$, $n = 1,2,\ldots$. The edges of $\mathcal{H}$ are defined as before, and the vertex $s_{\mathcal{H}}$, the class of the identity bimodule ${}_M M_M$, is distinguished. We have in fact four graphs, having as vertices classes of irreducible $N - N$, $N - M$, $M - N$ and $M - M$ bimodules respectively, and a typical picture could be the following.

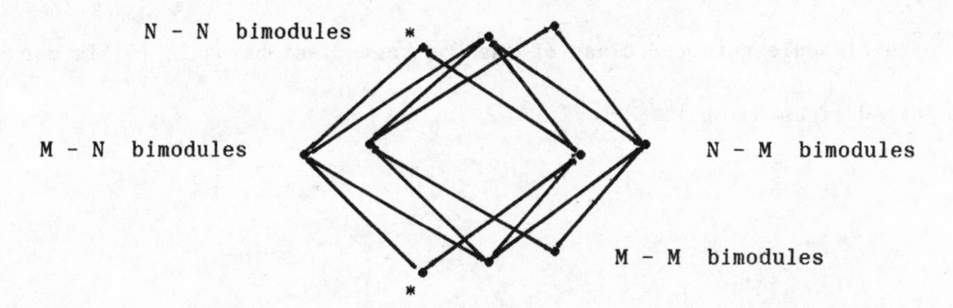

We compare now two ways of inducing an $N - N$ bimodule to an $M - M$ bimodule: inducing first on the right hand side or inducing first on the left hand side. Although the end result is the same, the decompositions into irreducible components at each stage yield, when compared, essential new information in the form of a scalar for each cell consisting of one edge in each of the four graphs. The corresponding map is called a connection.

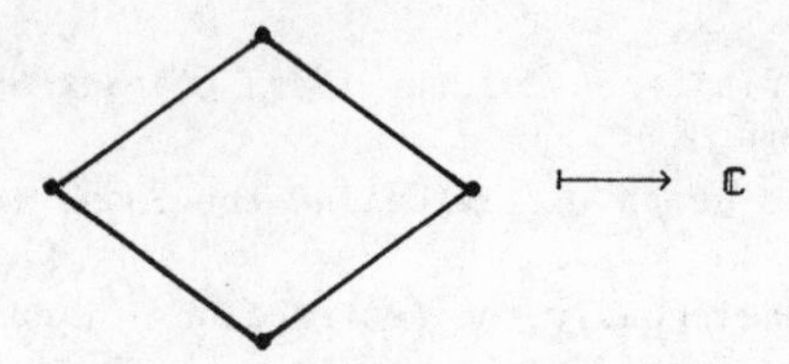

As an example, in the case when M is the crossed product of N by an outer action of a finite abelian group G the graphs depend only on the cardinality of G while the connection gives the pairing $G \times \hat{G} \to \mathbb{C}$. Because of the existence in our context of the contragredient of a bimodule, these graphs are two by two isomorphic, and the picture can be condensed to the two graphs $\mathfrak{G}$ and $\mathfrak{H}$. We have to record the involutive bijection τ, called the contragredient map, on the vertices of $\mathfrak{G} \cup \mathfrak{H}$, which maps the class of a bimodule onto the class of the contragredient bimodule (τ is marked with dotted lines below.)

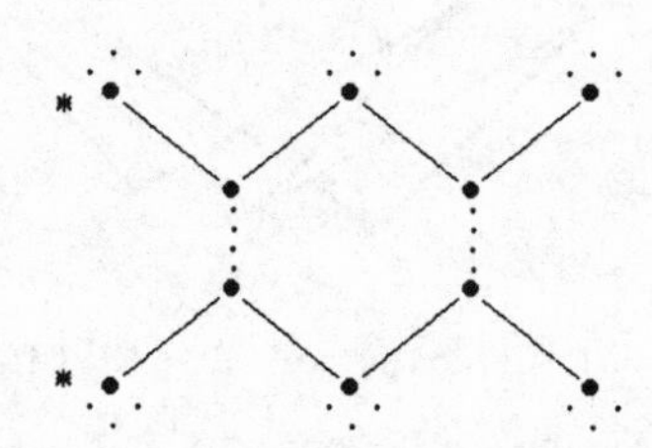

Choose now an irreducible bimodule from each equivalence class and a bimodule intertwiner for each edge of the graphs, as well as antiintertwiners corresponding to each pair of contragredient vertices; this is called a representation of $(\mathfrak{G}, \mathfrak{H}, \tau)$. A cell $c = (a_1, a_2, a_3, a_4)$ becomes now a quadruple of (oriented) edges of $\mathfrak{G} \cup \mathfrak{H}$ with the range y_i of a_i equal to the contragredient of the source x_{i+1} of a_{i+1} (indices mod 4).

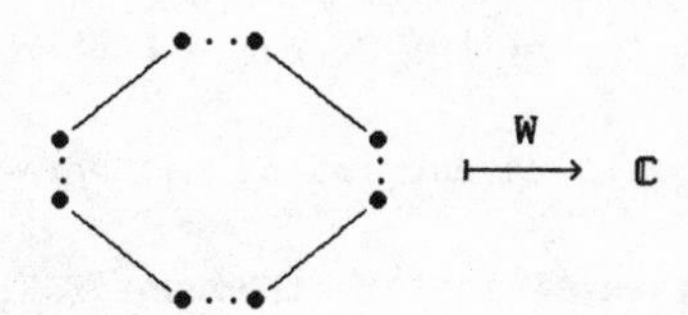

There is an i for which $x_i \in \mathcal{G}^{(0)}_{even}$; let ${}_N X_N$ be the representative of it. The composition of the intertwiners and antiintertwiners corresponding to the successive edges and pairs of vertices a_i, (y_i, x_{i+1}), a_{i+1}, (y_{i+1}, x_{i+2}), ..., a_{i+3}, (y_{i+3}, x_i) viewed as (anti)intertwiners of $N - N$ bimodules is an $N - N$ intertwiner of the irreducible bimodule ${}_N X_N$, and hence a scalar $W(c)$ (with suitable normalizations.) This way we obtain a complex valued map W, called a connection, on the cells of $\mathcal{G} \bigcup \mathcal{H}$; its equivalence class $[W]$ modulo the choice of the representation of $(\mathcal{G}, \mathcal{H}, \tau)$ is a conjugacy invariant of $N \subset M$. We call $(\mathcal{G}, \mathcal{H}, \tau, [W])$ the Galois invariant of $N \subset M$. The radius from $*_{\mathcal{G}}$ of the graph $\mathcal{G}$ is called the depth of $N \subset M$, and can be infinite even if the index $[M : N]$ is finite; however when $[M : N] < 4$ the depth is finite.

Define a function μ, called the Haar measure, on the vertices of $\mathcal{G} \bigcup \mathcal{H}$ with values in $\mathbb{R}^+$, by $\mu(x) = \dim({}_P X)$ if ${}_P X_Q$ is a bimodule representing x, where $P, Q \in \{M, N\}$. It is not hard to see that if Δ is the incidence matrix (or the laplacian) of $\mathcal{G} \bigcup \mathcal{H}$, then $\Delta \mu = \beta . \mu$, with

$\beta = [M : N]^{1/2}$, and $\mu(*_{\mathfrak{G}}) = \mu(*_{\mathfrak{H}}) = 1$. In the finite depth case, when $\mathfrak{G}$ and $\mathfrak{H}$ are finite, by the Perron - Frobenius theory μ is the only eigenvector with positive entries of Δ normalized at $*_{\mathfrak{G}}$ and $*_{\mathfrak{H}}$, and $\beta = \| \Delta \|$. But by an elementary theorem of Kronecker, if a matrix with natural entries has norm less than 2, then the norm must be $2 \cos(\pi/n)$, $n = 3,4,\ldots$. This yields the rigdity of the index $[M : N] = \beta^2$, since the same method shows that infinite graphs have Perron eigenvalues larger than or equal to 2.

THE INVARIANT

It is possible to axiomatize the Galois invariant of subfactors $N \subset M$ with finite depth. Any quadruple $(\mathfrak{G}, \mathfrak{H}, \tau, [W])$ satisfying the axioms (and in particular any Galois invariant) is called a coupling system. There are initialization axioms, describing the graphs and the connection around $*$, and then local and global axioms for the connection. A cell has 4 edges, and when 2 opposite vertices are fixed the connection W yields a complex matrix indexed with the two halves of the cell. The main local axiom states that for each choice of the 2 fixed vertices, this matrix is, up to a normalization, unitary. The connection W gives a transport for strings in $\mathfrak{G} \cup \mathfrak{H}$. Let (ξ_1,ξ_2) be an n-string on $\mathfrak{G}$ and let ν_1,ν_2 be paths on $\mathfrak{G}$

with even length k, source $*$ and ranges x_1 and respectively x_2. The transport $T_{\nu_1,\nu_2}((\xi_1,\xi_2))$ of (ξ_1,ξ_2) along (ν_1,ν_2) is a linear combination of pairs of paths (η_1,η_2) having length n, common range and sources x_1 and respectively x_2. The coefficient c of (η_1,η_2) is computed as follows. Consider the diagram

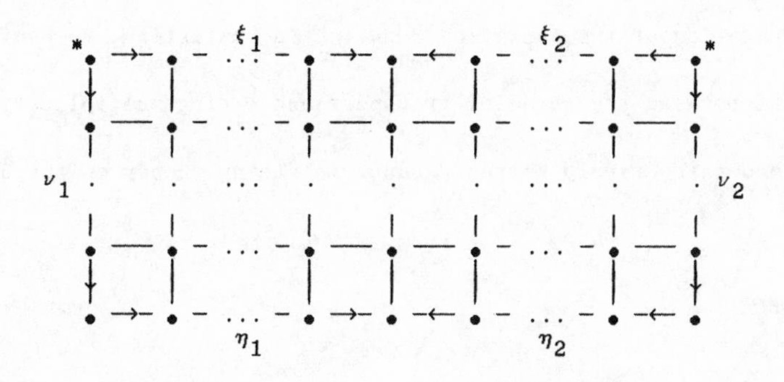

in which the space inside the contour is filled with cells of the coupling system $(\mathcal{G}, \mathcal{H}, \tau, [W])$; such a filling with cells having matching walls is called a surface. The contribution of each surface to c is the product

$$\prod W(\text{cell}_1) \cdot \prod W(\text{cell}_2)^{-}$$

where cell_1 ranges over the cells in the left half of the surface and cell_2 ranges over the cells in the right half. The coefficient c of (η_1,η_2) is obtained by summing up these products over all possible surfaces. If the length of ν_1 and ν_2 is odd, a similarly defined transport moves (ξ_1,ξ_2) into pairs of paths on the dual graph $\mathcal{H}$. The transport of strings on $\mathcal{H}$ is defined analogously. The main global axiom states that the transport T_{ν_1,ν_2} of any string is 0 unless $\nu_1 = \nu_2$, and $T_{\nu,\nu}$ depends only on the range x of ν, and will be denoted in the sequel by T_y, with $y = \tau(x)$. Thus the the transport of strings based at $*$ is a parallel transport, i.e. $\mathfrak{G}$ and $\mathfrak{H}$ are geometrically $*$-flat. Remark the formal similarity between the computation of the coefficients of the transport and the computation of the partition function in statistical mechanics on the one hand, and between the parallel transport and string multiplication and the action and respectively Witten product in string theory on the other hand.

THE MODEL

Starting from a finite depth coupling system $(\mathfrak{G}, \mathfrak{H}, \tau, [W])$ we construct a model of finite index subfactor of the elementary algebra R, called the string model.

For $n = 0,1,2,\ldots$ construct the string algebra A_n of the graph $\mathfrak{G}$ as follows. An n - string ρ on $\mathfrak{G}$ is a pair (ξ,η) of paths on $\mathfrak{G}$ which have source $*_{\mathfrak{G}}$, common range (which is called the range of ρ) and common length n (which is called the length of ρ.) The algebra A_n has linear basis n-strings and product defined by

$$(\xi_1,\eta_1)(\xi_2,\eta_2) = \delta(\eta_1,\xi_2)\ (\xi_1,\eta_2)$$

where δ is the Kronecker symbol. The adjoint $(\xi,\eta)^*$ of a string (ξ,η) is (η,ξ). The algebra A_n is a finite dimensional semisimple algebra over $\mathbb{C}$, or a finite dimensional C^*- algebra (called an FD-algebra in the sequel.) We map A_n into A_{n+1} by mapping an n-string (ξ,η) into the sum $\Sigma\ (\xi\circ\zeta,\eta\circ\zeta)$ over all edges ζ of $\mathcal{G}$ having source the range of (ξ,η); here $\circ$ denotes the composition of paths by juxtaposition. This map is an algebra homomorphism which is injective (here and in the sequel we assume that $\mathcal{G}$ has more than one edge) and thus yields an algebra $A_\infty = \bigcup A_n$.

For the graph $\mathcal{G}$, define the measure μ on the vertices as the unique eigenvector with nonnegative entries, normalized by $\mu(*_{\mathcal{G}}) = 1$, of the incidence matrix $\Delta_{\mathcal{G}}$ of $\mathcal{G}$, $\Delta_{\mathcal{G}}\ \mu = \beta.\mu$. Define now a positive trace $tr: A_n \to \mathbb{C}$, which is a linear functional given by

$$tr((\xi,\eta)) = \beta^{-n}\ \mu(x)\ \delta(\xi,\eta)$$

where (ξ,η) is a string of length n with range x. Since μ is an eigenvector of $\Delta_{\mathcal{G}}$, this trace is compatible with the inclusions $A_n \to A_{n+1}$ and thus gives a trace tr on A_∞. On the algebra A_∞ define in addition to the uniform norm $\|.\|$ inherited from the C^*-algebras A_n, the L^2-norm $\|.\|_2$ by $\|x\|_2 = (tr(x^*x))^{1/2}$. Complete now A_∞ with respect to $\|.\|$-bounded $\|.\|_2$-convergence to obtain a von Neumann algebra A (this type of completion would yield $L^\infty([0,1])$ from the algebra of polynomials on

[0,1].) Construct analogously the n-string algebras B_n on the graph $\mathfrak{H}$, and the string algebra B as the completion of $\bigcup B_n$. The algebras A and B, called the string algebras of the graphs $\mathfrak{G}$ and $\mathfrak{H}$, are isomorphic according to the theorem of Murray and von Neumann to the algebra R; in particular they retain no information about the graphs.

We now embed the string algebra A into the string algebra B by embedding for each n A_n into B_{n+1}; this yields a model of subfactor. We cannot, in general, use graph maps such as the embeddings of A_n into A_{n+1}; a more general class of maps, the cell maps, is needed (quite surprizingly any filtered map between locally finite dimensional algebras $\bigcup C_n$ and $\bigcup D_n$ is a cell map.) A linear map $\varphi: A_n \to B_{n+1}$ has the form $\varphi(\rho) = \Sigma\, c(\rho,\sigma)\, \sigma$ where ρ and σ are strings in A_n and respectively B_n, and the coefficients $c(\rho,\sigma) \in \mathbb{C}$. We use the connection W to define the coefficients $c(\rho,\eta)$ as follows. Let $\rho = (\rho_+,\rho_-)$ and $\sigma = (\sigma_+,\sigma_-)$, and construct a diagram

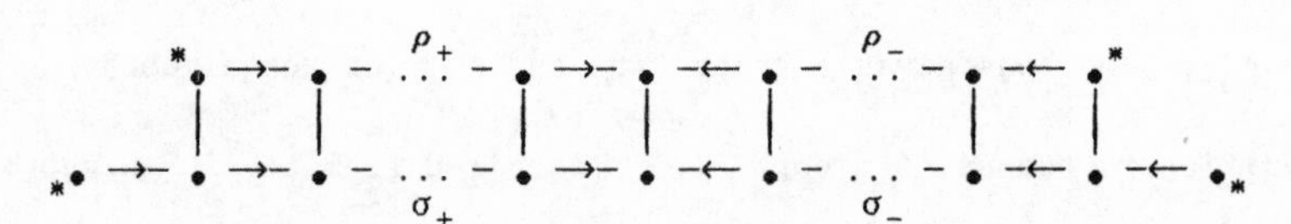

where the vertical lines are edges of $\mathfrak{G} \bigcup \mathfrak{H}$ such that each square is a cell; label the cells from left to right as $c_1, \ldots, c_n, c_n', \ldots, c_1'$. The coefficient $c(\rho,\sigma)$ is then the sum

$$\sum \ W(c_1)\ W(c_2)\ \ldots\ W(c_n)\ W(c'_n)^{-}\ \ldots\ W(c'_2)^{-}\ W(c'_1)^{-}$$

for all possible choices of vertical edges, where $^{-}$ denotes the complex conjugate. This is an instance of parallel transport.

The local unitarity of the connection W yields then the fact that the map φ is a $*$-homomorphism compatible with the inclusions $A_n \subset A_{n+1}$, $B_n \subset B_{n+1}$, which preserves the trace and thus gives a homomorphism $\varphi\colon A \to B$ of the string algebras. The inclusion $\varphi(A) \subset B$ is called the string model of subfactor on the coupling system $(\mathcal{G}, \mathcal{H}, \tau, [W])$. The main results in the sequel show that for subfactors with finite index, scalar centralizer and finite depth of the hyperfinite II_1 factor R, the Galois invariant and the string model are functors inverse to each other, and thus the coupling systems classify subfactors up to conjugacy.

The parallel transport given by the connection W is used to construct several morphisms of the string algebras A_n and B_n. The homomorphism $\varphi : A_n \to B_{n+1}$ can be extended, with essentially the same definition, to map $B_n \to A_{n+1}$. The square $\Gamma = \varphi\circ\varphi : A_n \to A_{n+2}$, $B_n \to B_{n+2}$ is called the canonical shift, and is a generalization of comultiplication in Hopf algebras. In a coupling system arising as a Galois invariant, the shift Γ is connected to Morita equivalence. The information in the coupling system can be recovered from the graph $\mathcal{G}$ together with the canonical shift Γ on its string algebras A_n, and the pair $(\mathcal{G}, \Gamma)$ can be axiomatized. With the parallel transport we also construct the mirrorings, which are involutive antihomomorphisms $\gamma_n : A_{2n} \longleftrightarrow A_{2n}$, $B_{2n} \longleftrightarrow B_{2n}$, $A_{2n+1} \longleftrightarrow B_{2n+1}$. For instance on A_{2n} the mirroring γ_n is an antiautomorphism which maps the n-string (ξ_1,ξ_2) into $\gamma_n((\xi_1,\xi_2)) = \Sigma\, c\, (\eta_1,\eta_2)$, where the coefficient c of the string (η_1,η_2) is computed as follows. Let x be the common range of η_1 and η_2. Then c is the coefficient of $(\eta_2{}^{\sim},\eta_1{}^{\sim})$ in the parallel transport $T_x((\xi_1,\xi_2))$ of (ξ_1,ξ_2) from $*$ to x, where $\sim$ denotes the path inverse. The mirrorings are generalizations of the coinvolution on Hopf algebras, and in the Galois invariant they are coming from the contragredient map on bimodules.

THE RANGE THEOREM

The following theorem shows that any coupling system with finite depth appears as the Galois invariant of a subfactor.

THEOREM Let $(\mathcal{G}, \mathcal{H}, \tau, [W])$ be a coupling system with finite depth. The string model built on $(\mathcal{G}, \mathcal{H}, \tau, [W])$ has Galois invariant isomorphic to $(\mathcal{G}, \mathcal{H}, \tau, [\overline{W}])$, where $\overline{W}$ is the connection complex conjugate to W.

The proof of this result begins by constructing open string bimodules as follows. Let x be an even vertex of the graph $\mathcal{G}$ and $n \geqq 0$. A $(*,x)$ open n-string is a pair (ξ,η) of paths on $\mathcal{G}$ of length n having common range such that ξ has source $*$ and η has source x. Let $A_n(x)$ be a linear space with basis $(*,x)$ open n-strings. An embedding of $A_n(x)$ into $A_{n+1}(x)$, an inner product on each $A_n(x)$ and finally the completion $A(x)$ of $\bigcup A_n(x)$ are defined the same way as for the string algebra A on $\mathcal{G}$. The n-string algebra A_n acts on $A_n(x)$ by

$$(\xi,\eta).(\xi_0,\eta_0) = \delta(\eta,\xi_0)\ (\xi,\eta_0)$$

This action is compatible with the inclusions and the trace and inner product for A_n and $A_n(x)$, and give an action of the string algebra A on $A(x)$ which gives $A(x)$ the structure of a left A-module. There is an action of A_n on $A_n(x)$ on the right defined as follows. Let ρ be an n-string on $\mathcal{G}$ and let $\sigma \in S_x^{(n)}$. Now use the connection W to (parallel) transport ρ from $*$ to x and with the result multiply the open string σ on the right. This gives $A_n(x)$ a right A_n-module structure and thus $A(x)$ becomes an $A - A$ bimodule. We have $A(*) = A$ as $A - A$ bimodules.

The main part of the proof of the theorem consists in showing that the bimodules constructed analogously for each vertex of $\mathcal{G} \bigcup \mathcal{H}$ are irreducible and that their intertwiners correspond to the edges of the graphs.

The irreducibility of the open string bimodules, which is an asymptotic property, is reduced using the mirrorings γ_n to an ergodic property of the random walk associated to the graphs coming from the Perron - Frobenius property.

THE CLASSIFICATION THEOREM

The string model algebras are by construction approximately finite dimensional. The following result shows that for approximately finite dimensional algebras this is the most general type of inclusion, in the finite depth case.

THEOREM Let N be a subfactor with finite index, scalar centralizer and finite depth of the elementary factor M. Then the subfactor $N \subset M$ is conjugate to the string model $A_{str} \subset B_{str}$ constructed on the conjugate $(\mathfrak{G}, \mathfrak{H}, \tau, [\overline{W}])$ of the Galois invariant $(\mathfrak{G}, \mathfrak{H}, \tau, [W])$ of $N \subset M$.

The proof starts with two constructions associated to the inclusion $N \subset M$. The upward construction is $N \subset M \subset \mathrm{End}({}_N M) \cong M \otimes_N M$ and the downward construction is a subfactor $P \subset N \subset M$ such that $M \cong \mathrm{End}({}_P N)$, the endomorphisms of N as a P-module. These constructions, called the basic constructions, were first systematically used by V. Jones. They can be iterated to give a sequence

$$\ldots \subset M_{-2} \subset M_{-1}=N \subset M_0=M \subset M_1 \subset M_2 \subset \ldots$$

The sequence $(M_n)_{n\geq -1}$, called the tower, is canonical. In the sequence $(M_n)_{n\leq 0}$, called the tunnel, the choice of any finite subtunnel $(M_k)_{n\leq k\leq 0}$ is unique only up to an inner automorphism of N.

The first part of the proof, algebraic in nature, is the encoding argument. It consists of showing that the inclusions of centralizers $M_{-n-1}' \cap M_{-1} \subset M_{-n-1}' \cap M_0$ are in fact opposite conjugate to the inclusions $A_n \subset B_{n+1}$ of string algebras on the Galois invariant of $N \subset M$.
This is done by first showing that $A_n \cong N' \cap M_{n-1}$ and $B_n \cong M' \cap M_n$.
For instance we have by the construction of the induction restriction graph $\mathcal{G}$

$$A_{2n} \cong \mathrm{End}((\mathrm{Res}_{N-N}\ \mathrm{Ind}_{M-N})^n({}_N N_N)) = \mathrm{End}({}_N(M^{\otimes n})_N)$$

with $M^{\otimes n} = M \otimes_N M \otimes_N \cdots \otimes_N M$ (n times). Further

$$\mathrm{End}({}_N(M^{\otimes n})_N) = N' \cap \mathrm{End}((M^{\otimes n})_N) = N' \cap M^{\otimes 2n} = N' \cap M_{2n-1}.$$

We use the canonical shift Γ to produce isomorphisms $A_n \cong M_{-1}' \cap M_{n-1} \cong M_1' \cap M_{n+1} \subset M_0' \cap M_{n+1} \cong B_{n+1}$ and show that this map corresponds to the cell homomorphism φ. Finally we use a mirroring (with respect to M_0) to show that $M_1' \cap M_{n+1} \subset M_0 \cap M_{n+1}$ is anticonjugate to $M_{-n-1}' \cap M_{-1} \subset M_{-n-1}' \cap M_0$.

The second part of the proof, which is analytic, is the spanning part. It consists of showing that the freedom of choice for any finite number of steps of the tunnel $M_{-n} \subset \ldots \subset M_{-1} \subset M_0$ can be used to insure that the towers of centralizers $(M_{-n}' \cap N)_{n \geqq 1}$ and $(M_{-n}' \cap M)_{n \geqq 1}$ span N and respectively M. By the first part of the proof, these centralizers are in fact a string model naturally built into the subfactor $N \subset M$. The freedom of choice for a finite number of steps of the tunnel is an inner automorphism of N; if instead we could use an inner automorphism of M then the spanning could be obtained by a classical hyperfiniteness argument. The latter case can be reduced to the former via a decoupling theorem, which states roughly that it is possible to choose the tunnel so that M_0 almost breaks into a tensor product $M_{-n} \otimes (M_{-n}' \cap M_0)$; this is analogue to an asymptotic abelianness argument in quantum field theory. The decoupling theorem is proved using the ergodic theory of the associated random walks on graphs, together with the study of central sequences in ultraproducts of the algebras. The main ingredients of the proof are the following. Here ω is a free ultrafilter on $\mathbb{N}$, and M^{ω} denotes the ω-ultraproduct algebra on M.

THEOREM Let $A \subset B \subset C$ be finite von Neumann algebras, with B approximately finite dimensional. Then

$$(A' \cap B^{\omega})' \cap C^{\omega} = A \vee (B' \cap C)^{\omega}.$$

This generalizes to paragroups the fact that outer automorphisms of the elementary algebra are outer on central sequences.

THEOREM Let $N \subset M$ be a subfactor with finite index, finite depth and scalar commutant of the elementary factor. Consider the tower of extensions $M_0 = N \subset M_1 = M \subset M_2 \subset M_3 \subset \ldots \subset M_\infty = \bigvee_k M_k$, and let $A = M' \cap M_\infty$.

Let E_M denote the conditional expectation onto the subalgebra M.

Let $\iota = \frac{1}{2} \| \mu \|_2^2$ where $\| \mu \|_2$ is the l^2-norm of the Haar measure μ.

a) The subfactor $M \otimes A$ of M_∞ has scalar relative commutant and index ι.

b) The subfactor $M' \cap N^\omega$ of $M' \cap M^\omega$ has scalar relative commutant and index ι.

c) Let $\tilde{M}$ and $\tilde{e}_1, \tilde{e}_2, \ldots$ be a subfactor and standard projections making the following tower standard

$$\tilde{M} \subset M \otimes A \subset M_\infty = \langle M \otimes A, \tilde{e}_1 \rangle \subset \langle M \otimes A, \tilde{e}_1, \tilde{e}_2 \rangle \subset \ldots.$$

Then the following tower is standard

$$M' \cap N^\omega \subset M' \cap M^\omega \subset \langle M' \cap M^\omega, \tilde{e}_1 \rangle \subset \langle M' \cap M^\omega, \tilde{e}_1, \tilde{e}_2 \rangle \subset \ldots.$$

d) Let $x \in M' \cap M^\omega$. Then

$$E_{M' \cap N^\omega}(x) = E_{\tilde{M}^\omega}(x) = \lim_{k \to \infty} E_{(M_{-k})^\omega}(x)$$

uniformly in $\|\cdot\|_2$-norm for $\|x\| \leqq 1$, where $\dots \subset M_{-2} \subset M_{-1} \subset M_0 \subset M_1$ is any standard tunnel.

Example. For the subfactors $N \subset M$ with graphs the Coxeter-Dynkin diagrams A_n the index is $[M : N] = 4 \cos^2 \frac{\pi}{n+1}$ while their central sequences have index

$$[M' \cap M^\omega : M' \cap N^\omega] = (n+1)(4 \sin^2 \frac{\pi}{n+1})^{-1}.$$

The algebra $\tilde{M} \subset M \otimes A$ is called a <u>core</u> for the subfactor $N \subset M$. Remark that in general if $N \subset M$ is as above, for a sequence $x = (x_n)_n \in M' \cap M^\omega$ the termwise conditional expectation $E_{N^\omega}(x) = (E_N(x_n))_n$ is not centralizing for M, which would be the case for cross products or fixed point algebras of group actions. Nevertheless, although the core is not even part of M, $E_{\tilde{M}^\omega}(x) = (E_{\tilde{M}}(x_n))_n$ is in $M' \cap N^\omega$. The core is unique up to an inner perturbation by a unitary in $M.A$, which does not change $E_{\tilde{M}^\omega}(x)$.

SUBFACTORS OF INDEX LESS THAN 4

After showing that the Galois invariant is faithful, we are left with the task of finding the possible coupling systems. While a complete description for all the values of the index is out of reach (e.g. all finite groups are in the range of the invariant,) it is possible to get a feeling of the general situation from the classification of subfactors with index less than 4, that is up to the first accumulation point of the index values. In this range all subfactors have automatically scalar centralizer and finite depth. The fact that the norm of the incidence matrix is less than 2 forces each graph to be a Coxeter - Dynkin diagram A_n, D_n or E_6, E_7, E_8. Conditions stemming from the biunitarity of the connection show that in this range the graphs $\mathcal{G}$ and $\mathcal{H}$ are isomorphic. Due to the simple structure of these graphs, it is not hard to find all the connections which satisfy the local axioms; up to isomorphism there is one for each diagram A_n or D_{2n+1}, there are two on each diagram D_{2n} differing by the contragredient map and there are two conjugate to each other for each diagram E_6, E_7, E_8. Finally the global axioms eliminate one connection for each D_n and the pair of connections on E_7. Thus there is one subfactor for each diagram A_n, one for each diagram D_{2n} and a pair of opposite conjugate but nonconjugate subfactors for each diagram E_6 and E_8.

PARAGROUPS

The coupling system has been defined as a Galois invariant containing as much information as possible, by using an analogue of the coupling between an abelian group and its dual. This approach is convenient for the exhaustion part of the classification properties: it is easier to find all groups in a given class by working with both the groups and their representations. For other purposes it may be useful to have the analogue of the group itself; in our context this is called a paragroup (para = standing for.)

A local description of the paragroup associated to an inclusion $N \subset M$ as before is the following. Let $\mathfrak{G}$ be the induction - restriction graph and τ^m the restriction to the even vertices of $\mathfrak{G}$ of the contragredient map τ. A macrocell consists of eight edges of $\mathfrak{G}$ as follows

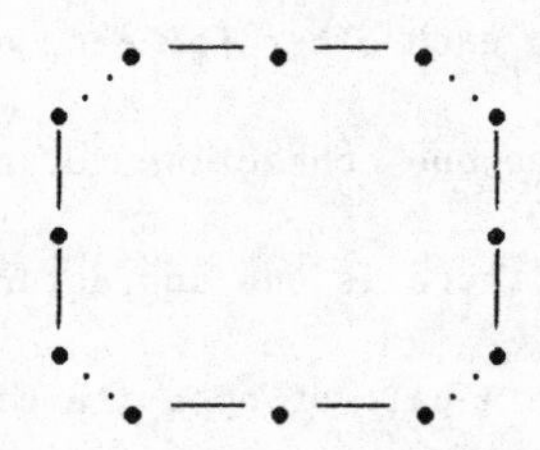

with endpoints alternatively equal and respectively contragredient to each other (contragredient vertices are connected by a dotted line in the figure.) To each such macrocell c we associate a scalar $W^m(c)$, with the map W^m called the macrocell connection. One way to construct W^m from the connection W of the coupling system $(\mathfrak{G}, \mathfrak{H}, \tau, W)$ is analogous to the computations for the parallel transport. We fill c with cells

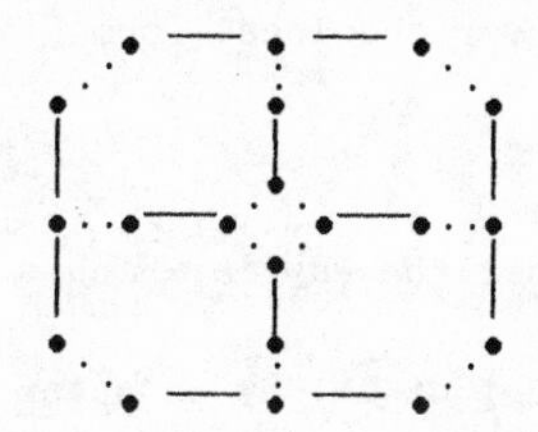

where the inner edges belong to the graph $\mathcal{H}$ and then take the product of W on the cells and sum over all ways of filling c. It is also possible to define W^m directly, in a way analogous to W. The triple $(\mathcal{G}, \tau^m, [W^m])$ forms a paragroup, and can be axiomatized in a manner similar to that of the coupling system $(\mathcal{G}, \mathcal{H}, \tau, [W])$. There is an analogous dual paragroup $(\mathcal{H}, \tau^m, [W^m])$ built on the restriction - induction graph $\mathcal{H}$, and the coupling system $(\mathcal{G}, \mathcal{H}, \tau, [W])$ plays the role of the Pontryagin - van Kampen duality coupling between these paragroups. The process of finding $\mathcal{H}$ and W from $(\mathcal{G}, \tau^m, [W^m])$ is the analogue of the representation theory for (finite) groups.

It is also possible to give a global description of a paragroup as follows. If ρ is an n-string on the graph $\mathcal{G}$ and x is an even vertex of $\mathcal{G}$ then the parallel transport T_x maps ρ into a linear combination of n-strings on $\mathcal{G}$ with source x. Remark that in the construction of T_x we only need to know W^m. The pair $(\mathcal{G}, (T_x))$ where $x \in \mathcal{G}^{(0)}_{even}$ is the global form of a paragroup, can be completely axiomatized and contains all the information needed to recover the local form $(\mathcal{G}, \tau^m, [W^m])$.

As an example, consider the way in which a group appears as a paragroup. Let M be the cross product of N by an outer action of the finite group G. Then the induction - restriction graph $\mathcal{G}$ has a single odd vertex $**$ and its even vertices are indexed by G with $*_{\mathcal{G}}$ at the unit 1_G. There is a single edge labeled g between the vertex g and $**$ for each $g \in G$. The contragredient map τ^m connects the vertices g and g^{-1}. For $G = \mathbb{Z}_3$ the picture is the following.

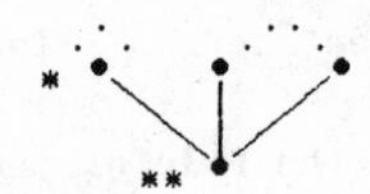

A macrocell c is labeled as follows.

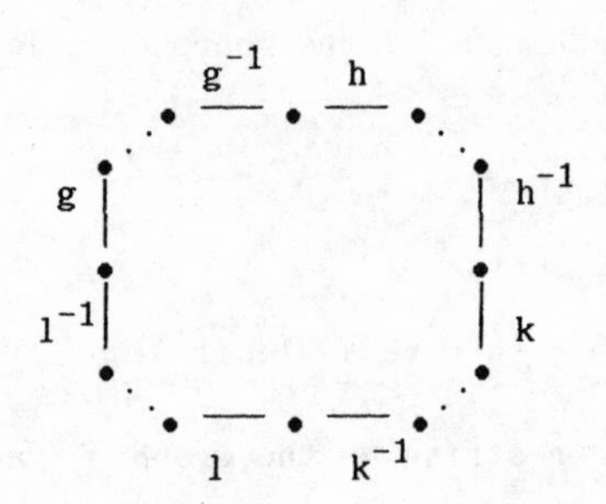

Then $W^m(c)$ is 1 if $ghkl = 1$ and 0 otherwise, and thus yields the multiplication table of G. The parallel transport of a 2-string $\rho = ((1,g),(1,g))$ to the vertex h is the string $T_h(\rho) = ((h,hg),(h,hg))$. The picture for a finite quantum group is analogous, with the parallel transport yielding the comultiplication.

QUANTIZED DYNAMICAL SYSTEMS

In classical mechanics, a dynamical system consists of a group G acting on a measure space (X,μ). Instead of the space X one can work with the algebra $A = L^{\infty}(X,\mu)$ of bounded measurable functions on X, on which the group acts by an automorphic action $\alpha: G \longrightarrow \mathrm{Aut}\, A$. In a quantum mechanical setting, the algebra $L^{\infty}(X,\mu)$ is replaced by the noncommutative algebra $B(H)$ of bounded operators on a Hilbert space H, or by a weakly closed *-subalgebra A of $B(H)$. The quantized dynamical system consists of an automorphic action $\alpha: G \longrightarrow \mathrm{Aut}\, A$ of the group G on the noncommutative von Neumann algebra A.

There is growing evidence that the quantization of the space X should be accompanied by a corresponding quantization of the group G, since the noncommutative algebra A has more symmetry than its automorphism group can detect. Consider, for instance, a crossed product algebra $A = B \rtimes G$ of an algebra B by a group action $\alpha: G \longrightarrow \mathrm{Aut}\, B$. If the group G is abelian, then there is an action $\alpha\hat{}: G\hat{} \longrightarrow \mathrm{Aut}\, A$, where $G\hat{}$ is the dual of G (in the Pontryagin - van Kampen sense), with $\hat{\alpha}_{\sigma}(\Sigma\, x_g g) = \Sigma\, \sigma(g) x_g g$ for $g \in G$,

$\sigma \in G^{\wedge}$, $x_g \in B$, and the algebra B consists of the fixed points of the dual action $\alpha^{\wedge}$. If the group G is compact nonabelian, then to an irreducible representation $\sigma: G \to Mat_n\mathbb{C}$ there corresponds a homomorphism $\alpha^{\wedge}_{\sigma}: A \to A \otimes Mat_n\mathbb{C}: \Sigma\, x_g g \to \Sigma\, x_g g \otimes \sigma(g)$. This is an action of the dual $G^{\wedge}$ of G, with $G^{\wedge}$ viewed as a quantum group [Dr] (Hopf algebra, ring group [Ka], Kac algebra [Ta,ES]), and its fixed point set $\{y \in A;\ \alpha^{\wedge}_{\sigma}(y) = y \otimes 1,\ \sigma \in \mathrm{Irr}\, G\}$ is again the algebra B. Homomorphisms $A \to A \otimes Mat_n\mathbb{C}$ such as $\alpha^{\wedge}_{\sigma}$ should be viewed as quantized symmetries of the algebra A. The homomorphism $\alpha^{\wedge}_{\sigma}$ is generalized invertible: if $\sigma-$ is the conjugate of σ, then $(\alpha^{\wedge}_{\sigma-} \otimes id_n) \circ \alpha^{\wedge}_{\sigma}$ contains the identity id_A under the projection $\frac{1}{n} \sum_{i,j} e_{ij} \otimes e_{ij} \in Mat_n\mathbb{C} \otimes Mat_n\mathbb{C}$, where e_{ij} are the matrix units of $Mat_n\mathbb{C}$. In certain situations, such as the quantum Yang-Baxter equations in statistical mechanics, one has to consider similarly defined actions of quantum groups which are neither groups nor group duals [Dr]. Harmonic analysis concepts such as the Haar measure, the Fourier transform, the Pontryagin duality theorem and the Takesaki duality for crossed products extend naturally to the quantum group setting.

What is then the suitable quantization of the group concept which should accompany the quantization of the space? Ideally, like the finite abelian groups, the category should be closed under quotients by subobjects and under duality, with the bidual of an object isomorphic to the object. Quantum groups are closed under duality, but only quotients by normal quantum subgroups are again quantum groups. Separable field extensions can be viewed as indexed by a quotient G/H, where the group G corresponds to the normal extension and H is a subgroup of G. This category is closed under quotients.

A natural answer is to let the quantization of spaces to algebras determine the quantization of groups via Galois theory. The quantized groups should appear as those (quantized) symmetries of an algebra which leave the subalgebra invariant. If a group G acts freely (i.e. outerly) on an algebra A, the position of the algebra A inside the crossed product $B = A \times G$ gives back the group G. If $B_1 = A_1 \times G_1$ is another such crossed product and there is an algebra isomorphism $\theta: B \rightarrow B_1$ with $\theta(A) = A_1$ (in which case we say that A has the same position in B as A_1 in B_1), then the groups G and G_1 are isomorphic. A direct way to see this in the case when A is a factor, i.e. has scalar center, is the following. The unique decomposition of B as an $A - A$ - bimodule into irreducible subbimodules is $_AB_A = \bigoplus_{g\in G} {}_AAg_A$, and the tensor product $_AAg_A \otimes_A {}_AAh_A$ is isomorphic as an $A - A$ - bimodule to $_AAgh_A$. Thus G is obtained as the group of equivalence classes of irreducible subbimodules of $_AB_A$ with composition given by the tensor product over A. Similarly, the group-like object which we introduced appears as the position invariant for a subalgebra inside an algebra. In fact it is convenient to work with both the group - like object and its dual, which form a coupling system. Homomorphisms of a von Neumann algebra A into an ampliation $A \otimes Mat_n$ (i.e. quantized symmetries, as described above) correspond to bimodules ${}_A X_A$. Thus the invariants introduced in terms of bimodules can alternatively be given in terms of quantized symmetries, and the paragroup and coupling system are a way to organize these quantized symmetries in a group - like form.

In these notes we work under the finite index assumption, and obtain the quantized analogues of finite groups. The fact that even in this context the theory is expressed naturally in geometric terms suggests that an extension of these methods and results to a differentiable context could have very interesting consequences in quantum physics and geometry.

APPENDIX A: AXIOMS FOR COUPLING SYSTEMS AND PARAGROUPS

a) NOTATION

Let $\mathcal{G}$ be a graph, with vertices $\mathcal{G}^{(0)}$ and edges $\mathcal{G}^{(1)}$. For an edge e let $s(e)$ and $r(e) \in \mathcal{G}^{(0)}$ denote its source and range. We shall consider an unoriented graph $\mathcal{G}$ as an oriented graph endowed with an involution $e \mapsto e^{\sim}$ on $\mathcal{G}^{(1)}$, with $s(e^{\sim}) = r(e)$. In the sequel $\mathcal{G}$ is an unoriented graph.
A <u>path</u> ξ is an n-uple of edges $(e_1, e_2, \ldots, e_n)$ with $s(e_{i+1}) = r(e_i)$. The path ξ has source $s(\xi) = s(e_1)$, range $r(\xi) = r(e_n)$ and length $|\xi| = n$. A path ξ of length 0 is a vertex x, with $s(\xi) = r(\xi) = x$. The inverse $\xi^{\sim}$ of ξ is $(e_n^{\sim}, e_{n-1}^{\sim}, \ldots, e_1^{\sim})$ and the composition of the paths $\xi = (e_1, \ldots, e_n)$ and $\zeta = (f_1, \ldots, f_m)$ with $r(\xi) = s(\zeta)$ is the path $\xi \circ \zeta = (e_1, \ldots, e_n, f_1, \ldots, f_m)$. Let $\mathrm{Path}^{(n)}$ denote the set of paths ξ with $|\xi| = n$, and let $\mathrm{Path}_x^{(n)} = \{\xi \in \mathrm{Path}^{(n)};\ s(\xi) = x\}$ and $\mathrm{Path}_{x,y}^{(n)} = \{\xi \in \mathrm{Path}_x^{(n)};\ r(\xi) = y\}$. The cardinal of $\mathrm{Path}_{x,y}^{(n)}$ is denoted by $p(x,y;n)$.

b) MEASURE GRAPHS

A <u>standard</u> <u>measure</u> <u>graph</u> is a couple $(\mathcal{G}, \mu)$.

$\mathcal{G}$ is a graph which is connected, locally finite and unoriented, with the number of edges adjacent to a vertex uniformly bounded.

The graph $\mathcal{G}$ has a distinguished vertex $* = *_{\mathcal{G}}$ adjacent to only one edge and has no cycles of odd length; its vertices $\mathcal{G}^{(0)}$ are divided into two classes, $\mathcal{G}^{(0)}_{\text{even}}$ and $\mathcal{G}^{(0)}_{\text{odd}}$, according to the parity of their distance to $*$.

The measure μ is a map $\mathcal{G}^{(0)} \longrightarrow \mathbb{R}^+$ with $\mu(*) = 1$. Let Δ denote the Laplacian operator (having matrix the incidence matrix) of $\mathcal{G}$. The measure must satisfy the following axiom.

(H) Harmonicity

The measure μ is an eigenvector for the Laplacian

$$\Delta \mu = \beta \mu \qquad \text{with } \beta \in \mathbb{R}^+.$$

A measure with this property will be called a Haar measure on the graph $\mathcal{G}$.

REMARKS

We conjecture that for infinite graphs, the following axiom must be also satisfied

(PE) Path ergodicity

$$\lim_{k\to\infty} \sum_e \left| \beta^{-k} p(*,x;k)\, \mu(y) - \beta^{-k-1} p(*,y;k+1)\, \mu(x) \right| = 0$$

where $e \in \mathcal{G}^{(1)}$ and $x = s(e)$, $y = r(e)$.

For finite graphs the condition (H) determines the measure μ completely, $\beta = \| \Delta \|$ and (PE) is implied by (H) via the Perron - Frobenius theory. For infinite graphs the range of the eigenvalue β for harmonic measures is $[\ \|\Delta\|, +\infty\ [$, and it is an open question whether (PE) implies that $\beta = \|\Delta\|$.

In general if n is the supremum of the number of edges with the same source, then the operator norm $\|\Delta\|$ satisfies $n^{1/2} \leqq \|\Delta\| \leqq n$ and $\|\Delta_{\mathcal{G}}\| = \sup_n \|\Delta_{\mathcal{G}_n}\|$ for an ascending family $\mathcal{G}_n$ of subgraphs of $\mathcal{G}$ with union $\mathcal{G}$.

For finite graphs, Hoffman and Shearer have shown that the accumulation points of the range of the eigenvalue β are $\{r_n \mid n = 1,2,\ldots\} \cup [r_{cr}, +\infty[$, where $r_n = \alpha^{1/2} + \alpha^{-1/2}$ with α the positive root of the polynomial $x^{n+1} - 1 - x - \ldots - x^{n-1}$ and $r_{cr} = \varphi^{1/2} + \varphi^{-1/2}$, with φ the golden ratio. ($r_1 = 2$, $r_2 \cong 2.019$, $r_{cr} \cong 2.058$.) The lowest eigenvalue $\beta > 2$ is $\cong 2.007$.

The measure μ determines a stationary random walk on the graph $\mathcal{G}$, with initial point $*$ and transition probabilities proportional to μ. The condition (H) is known in statistical mechanics as detailed balance, and means that the transition probabilities for the reverse random walk are asymptotically proportional to μ; alternatively, the probability of arriving in k steps to a vertex x on a path ξ does not depend on ξ. The condition (PE) is equivalent to the fact that if $P_k(e)$ is the probability that the random walk uses the unoriented edge e from the step k to $k+1$, then $\|P_{k+1} - P_k\| \to 0$.

c) LOCAL COUPLING SYSTEMS

A <u>local</u> <u>coupling</u> <u>system</u> is a quadruple $(\mathfrak{G}, \mathfrak{H}, \tau, W)$.

$\mathfrak{G} = (\mathfrak{G}, \mu)$ and $\mathfrak{H} = (\mathfrak{H}, \mu)$ are disjoint standard measure graphs, with $\|\Delta_{\mathfrak{G}}\| = \|\Delta_{\mathfrak{H}}\|$. Let $**$ be the vertex in each graph adjacent to $*$.

The <u>contragredient</u> map τ is an involution of $\mathfrak{G}^{(0)} \cup \mathfrak{H}^{(0)}$ which preserves the measure, $\mu \circ \tau = \mu$, and satisfies the following conditions.

(CI) <u>Initialization</u>

$$\tau(*_{\mathfrak{G}}) = *_{\mathfrak{G}}$$

$$\tau(*_{\mathfrak{H}}) = *_{\mathfrak{H}}$$

$$\tau(**_{\mathfrak{G}}) = **_{\mathfrak{H}}.$$

(CP) <u>Parity</u>

$$\tau(\mathfrak{G}^{(0)}_{\text{even}}) = \mathfrak{G}^{(0)}_{\text{even}}$$

$$\tau(\mathfrak{H}^{(0)}_{\text{even}}) = \mathfrak{H}^{(0)}_{\text{even}}$$

$$\tau(\mathfrak{G}^{(0)}_{\text{odd}}) = \mathfrak{H}^{(0)}_{\text{odd}}.$$

The <u>connection</u> W is a map which associates to any <u>cell</u> (a_1, a_2, a_3, a_4) consisting of four oriented edges a_1, a_2, a_3, a_4 of $\mathfrak{G} \cup \mathfrak{H}$ with $s(a_{i+1}) = \tau(r(a_i))$ $i = 1, \ldots, 4$; $a_5 = a_1$, a number, the <u>energy</u> $W(a_1, a_2, a_3, a_4) \in \mathbb{C}$, having the following properties.

(WI) <u>Inversion</u> <u>symmetry</u>

$$W(a_4\tilde{}, a_3\tilde{}, a_2\tilde{}, a_1\tilde{}) = W(a_1, a_2, a_3, a_4)$$

where $\sim$ denotes the edge inverse.

(WR) Rotation symmetry

$$W(a_2,a_3,a_4,a_1) = W(a_1,a_2,a_3,a_4)^-$$

where $^-$ denotes complex conjugation.

(WU) Bi-unitarity

For any vertices $x,y \in \mathcal{G}^{(0)} \bigcup \mathcal{H}^{(0)}$ the matrix $(U_{i,j})$ is unitary, where the indices are pairs of edges $i = (a_1,a_2)$, $j = (a_3,a_4)$, with edges $a_1,a_2,a_3,a_4 \in \mathcal{G}^{(1)} \bigcup \mathcal{H}^{(1)}$, $s(a_i) = \tau(r(a_{i-1})) = x_i$, $i = 1,\ldots,4$, $a_0 = a_4$, having fixed ends $x_1 = x$, $x_3 = y$, and U is a renormalization of the connection W

$$U_{i,j} = \mu(x_1)^{-1/4}\mu(x_2)^{1/4}\mu(x_3)^{-1/4}\mu(x_4)^{1/4}\ W(a_1,a_2,a_3,a_4).$$

The empty matrix is considered unitary.

d) CELL CALCULUS

We extend the map W from cells to a map defined on contours as follows. A contour is a quadruple $(\xi_1,\xi_2,\xi_3,\xi_4)$ with $\xi_1,\xi_2,\xi_3,\xi_4 \in \text{Path}(\mathcal{G} \bigcup \mathcal{H})$ having $|\xi_i| = |\xi_{i+2}|$, $i = 1,2$ and $s(\xi_{i+1}) = \tau(r(\xi_i))$, $i = 1,\ldots,4$ where $\xi_5 = \xi_1$.

A surface s is a family of cells

$$(c(i,j)) = ((c(i,j)_1,c(i,j)_2,c(i,j)_3,c(i,j)_4))$$

$i = 1,\ldots,m$; $j = 1,\ldots,n$, having matching walls

$$c(i+1,j)_4 = c(i,j)_2^{\sim}$$

$$c(i,j+1)_1 = c(i,j)_3^{\sim}.$$

The boundary ∂s of the surface s is the contour $(\xi_1,\xi_2,\xi_3,\xi_4)$ where

$$\xi_1 = c(n,1)_1 \circ c(n-1,1)_1 \circ \dots \circ c(1,1)_1$$

$$\xi_2 = c(n,m)_2 \circ c(n,m-1)_2 \circ \dots \circ c(n,1)_2$$

$$\xi_3 = c(1,m)_3 \circ c(2,m)_3 \circ \dots \circ c(n,m)_3$$

$$\xi_4 = c(1,1)_4 \circ c(1,2)_4 \circ \dots \circ c(1,m)_4.$$

For the surface s, the <u>energy</u> $W(s)$ is the integral product

$$W(s) = \prod_{i,j} W(c(i,j)) \in \mathbb{C}.$$

Finally, for a contour c, the <u>energy</u> $W(c)$ is the sum

$$W(c) = \sum_s W(s)$$

over all surfaces s with boundary c.

If the contour c is degenerate, say $|\xi_1| = 0$, then we let $W(c) = \delta(\xi_2, \xi_4^{\sim})$; similarly, if $|\xi_2| = 0$, then $W(c) = \delta(\xi_1, \xi_3^{\sim})$. It is convenient to extend W to arbitrary quadruples of paths, by letting $W(\xi_1,\xi_2,\xi_3,\xi_4) = 0$ if $(\xi_1,\xi_2,\xi_3,\xi_4)$ is not a contour.

REMARK

There is a striking formal similarity between the computation of the energy of a contour and the expression of the partition function Z in string theory, where $Z = \sum_{\text{surfaces}} \exp(-S)$, and the action S is given by $S = \int_{\text{surface}} I(X,g)$, with I a form in the coordinates X and the induced metric g of the surface.

e) GLOBAL COUPLING SYSTEMS

The local coupling system $(\mathcal{G},\mathcal{H},\tau,W)$ is a <u>global coupling system</u> if the following two global axioms are satisfied.

(PT) The <u>parallel transport</u> axiom.

$$W(\xi_1,\xi_2,\xi_3,\xi_4) = \delta(\xi_1,\xi_3^{\sim})\ \delta(\xi_2,\xi_4^{\sim}).$$

for any contour with $s(\xi_i) = r(\xi_i) = * \in \{*_{\mathcal{G}},*_{\mathcal{H}}\}$, i=1,2,3,4.

(GC) The <u>global contragredient</u> axiom.

For any $x \in \mathcal{G}^{(0)} \cup \mathcal{H}^{(0)}$ there is a contour $c = (\xi_1,\xi_2,\xi_3,\xi_4)$, with $W(c) \neq 0$ such that $s(\xi_1) = s(\xi_3) = *$, $s(\xi_2) = x$, $s(\xi_4) = \tau(x)$.

This axiom takes care of the compatibility between the contragredient map and the cells.

REMARKS

The connection W defines, for paths ξ, η with $|\xi| = |\eta|$, a transport

$$T_{\xi,\eta}: X_{x,y} \to X_{z,t}$$

where $x = \tau(s(\xi))$, $y = \tau(s(\eta))$, $z = \tau(r(\xi))$, $t = \tau(r(\eta))$ and $X_{a,b}$ is the space of formal linear combinations of paths ρ with $s(\rho) = a$, $r(\rho) = b$, by

$$T_{\xi,\eta}(\rho) = \sum_{\gamma} c(\rho,\gamma)\ \gamma$$

with $c(\rho,\gamma) = W(\rho,\eta,\gamma^{\sim},\xi^{\sim}) \in \mathbb{C}$. It can be shown that the parallel transport axiom is equivalent to the fact that for any paths ξ,η with $s(\xi) = s(\eta) = *$ we have $T_{\xi,\eta} = 0$ unless $\xi = \eta$ in which case $T_{\xi,\xi}$ depends only on $r(\xi)$, i.e. the transport from $*$ is a parallel transport.

The axioms (PT) and (GT) are independent of the local axioms and are independent of each other. It can be shown that for the axiom (PT) needs to be checked only on contours with edges ξ_i monotonous, i.e. with length twice the distance from $*$ to the midpoint; a finite graph has only finitely many such contours.

f) COUPLING SYSTEMS

Let $(\mathcal{G},\mathcal{H},\tau,W)$ be a local coupling system. A perturbation u of the connection W consists of a complex unitary matrix $(u(a,b))_{a,b}$ associated to each pair of adjacent vertices $x,y \in \mathcal{G}^{(0)} \cup \mathcal{H}^{(0)}$, where $a,b \in \mathrm{Path}^{(1)}_{x,y}$, with

$$u(a^{\sim},b^{\sim}) = u(a,b)^{-}$$

for any a,b.

The perturbed connection is $W^{\#}$, where

$$W^{\#}(b_1,b_2,b_3,b_4) = \sum_{a_1,a_2,a_3,a_4} u(b_1,a_1)\, \overline{u(b_2,a_2)}\, u(b_3,a_3)\, \overline{u(b_4,a_4)}\, W(a_1,a_2,a_3,a_4) .$$

Here a_1,a_2,a_3,a_4 are edges with $s(a_i) = s(b_i)$, $r(a_i) = r(b_i)$, $i = 1,2,3,4$

The local coupling systems $(\mathcal{G},\mathcal{H},\tau,W)$ and $(\mathcal{G},\mathcal{H},\tau,W^{\#})$ are called equivalent. The global properties (PT) and (GC) are preserved under equivalence.

A <u>coupling system</u> $(\mathcal{G},\mathcal{H},\tau,[W])$ is the equivalence class of a global coupling system $(\mathcal{G},\mathcal{H},\tau,W)$ under perturbation. The coupling systems $(\mathcal{G},\mathcal{H},\tau,[W])$ and $(\mathcal{G}',\mathcal{H}',\tau',[W'])$ are <u>isomorphic</u> if there is a perturbation $W^{\#}$ of W and a graph isomorphism $\theta\colon \mathcal{G} \longrightarrow \mathcal{G}'$, $\mathcal{H} \longrightarrow \mathcal{H}'$ with $\mu' \circ \theta = \mu$, $\tau' \circ \theta = \theta \circ \tau$ and $W' \circ \theta = W^{\#}$.

APPENDIX B. THE COUPLING SYSTEMS FOR INDEX LESS THAN 4

THEOREM The coupling systems $(\mathfrak{G},\mathfrak{H},\tau,[W])$ of dimension less than 4 are described below.

The corresponding paragroups have graph $(\mathfrak{G},\tau_{even},[W_{macrocell}])$, where τ_{even} is the restriction of the contragredient map τ to the even vertices of $\mathfrak{G}$, and the macrocell connection $W_{macrocell}$ is computed from the connection W.

The coupled graphs

The vertices of the graphs $\mathfrak{G}$ and $\mathfrak{H}$ are labeled $1,\ldots,n$ and respectively $1',\ldots,n'$, with 1 and 1' the distinguished vertices. In dimension less than 4, the dual graphs $\mathfrak{G}$ and $\mathfrak{H}$ are isomorphic to each other. A vertex and its contragredient are joined by a dotted line.

Remark that in the graphs $\mathbb{D}_n$ for $n = 4k$ the endpoints $n-1$ and n are contragredient to each other while for $n = 4k+2$ they are selfcontragredient.

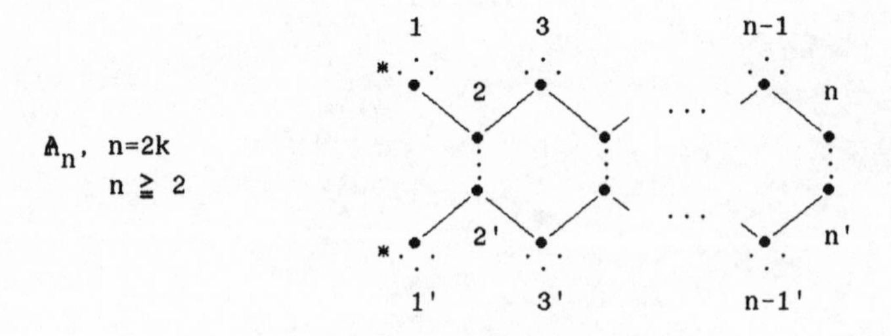

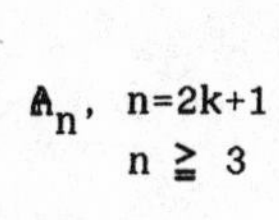

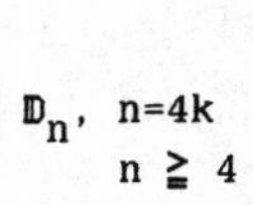

$\mathbb{D}_n$, n=4k+2
n ≧ 6

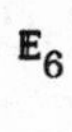

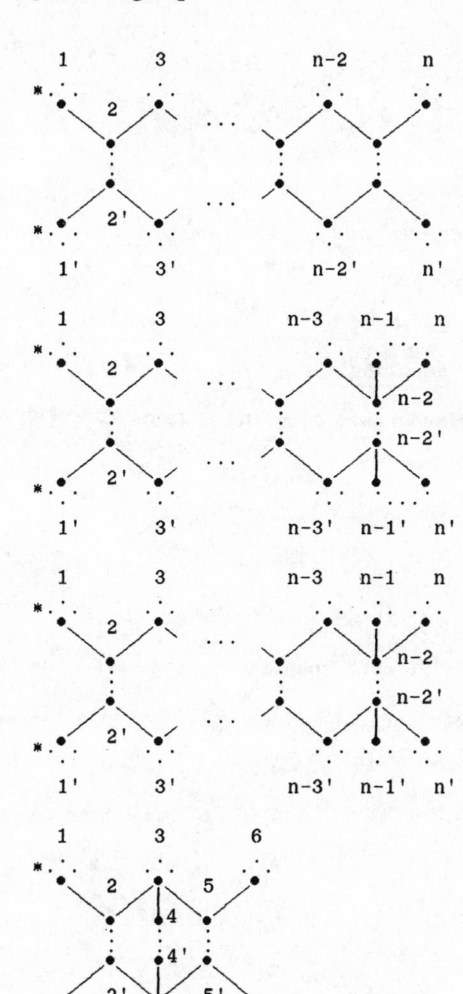

$\mathbb{E}_8$

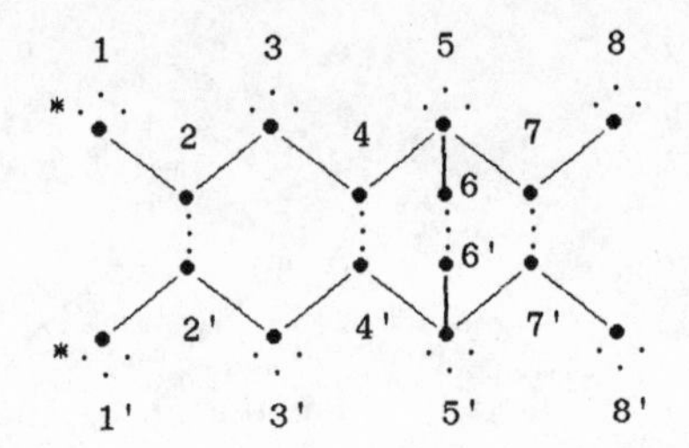

There is one coupling system for each of the graphs $\mathbb{A}_n$ and $\mathbb{D}_{2n}$, and there are two nonisomorphic coupling systems for each of the graphs $\mathbb{E}_6$ and $\mathbb{E}_8$.

The dimension

The dimension of each coupling system (and thus the index of the corresponding subfactor) is $4 \cos^2 \frac{\pi}{m}$, where m is the Coxeter number of the diagrams.

$$\mathbb{A}_n: \quad m = n + 1$$

$$\mathbb{D}_n: \quad m = 2n - 2$$

$$\mathbb{E}_6: \quad m = 12$$

$$\mathbb{E}_8: \quad m = 30$$

The Haar weight

Let $\alpha = \frac{\pi}{m}$ with m the Coxeter number of the graph. The Haar weight μ on the vertices labeled $1,2,\ldots;1',2',\ldots$ is given by the following expressions.

$\mathbb{A}_n$

$\mu(k) = \mu(k') = s(k)/s(1), \qquad k = 1,\dots,n.$

$\mathbb{D}_n$

$\mu(k) = \mu(k') = s(k)/s(1), \qquad k = 1,\dots,n-2,$

$\mu(n-1) = \mu((n-1)') = \mu(n) = \mu(n') = 1/2.$

$\mathbb{E}_n$

$\mu(k) = \mu(k') = s(k)/s(1), \qquad k = 1,\dots,n-3,$

$\mu(n-2) = \mu((n-2)') = s(n-2)/s(2),$

$\mu(n-1) = \mu((n-1)') = (s(n-2)s(2))/(s(3)s(1)),$

$\mu(n) = \mu(n') = s(n-2)/s(3).$

The connection

We specify a cell by the vertices (k_1,k_2,k_3,k_4) which are the sources of its edges (e_1,e_2,e_3,e_4), where $k_i \in \{1,\dots,n,1',\dots,n'\}$.
We let $\overline{j} = j'$ and $\overline{j'} = j$ for $j \in \{1,\dots,n\}$.

The cells are given by the following expression, common to all graphs $\mathbb{A}_n$, $\mathbb{D}_n$, $\mathbb{E}_n$.

Let m be the Coxeter number, and $\varepsilon = -i \exp\left(\frac{\pi i}{2m}\right)$, $i = \sqrt{-1}$.

The curvature of the cell $c = (k_1,k_2,k_3,k_4)$ is $R(c) = \left(\frac{\mu(k_1)\,\mu(k_3)}{\mu(k_2)\,\mu(k_4)}\right)^{1/2}$.

On the cell $c = (k_1,k_2,k_3,k_4)$ the connecion W takes then the value

$$W(c) = \varepsilon\, R(c)\, \delta(k_1,\overline{k_3}) + \overline{\varepsilon}\, R(c)^{-1}\, \delta(k_2,\overline{k_4}).$$

The above formulae give two connections corresponding to the two choices of $\sqrt{-1}$.

For the graphs $\mathbb{A}_n$, the two connections are equivalent. There is a connection, equivalent to these, with $W(c) \in \mathbb{R}$ for all cells c.

For the graphs $\mathbb{D}_n$ the two connections are nonequivalent but they are isomorphic using a symmetry $n-1 \leftrightarrow n$ in the graph $\mathcal{G}$. On the macrocells, these connections are equivalent, and there is an equivalent macrocell connection with real values.

For the graphs $\mathbb{E}_n$, the two connections are nonisomorphic and correspond to nonconjugate subfactors; changing from a subfactor $N \subset M$ to the opposite $N^{op} \subset M^{op}$ switches between these connections.

□

The subfactors with diagram $\mathbb{A}_n$ of the hyperfinite II_1 factor R are isomorphic to the subfactors generated by the projections e_n introduced by Vaughan Jones, and are refered to henceforth as the Jones factors.

For the integer values 1, 2 and 3 in the above dimension range the paragroups and coupling systems are also given by the following familiar constructions.

In dimension 1, 2 and 3 the paragroups with diagrams $\mathbb{A}_2$, $\mathbb{A}_3$ and respectively $\mathbb{D}_4$ correspond to the groups $\mathbb{Z}_1$, $\mathbb{Z}_2$ and respectively $\mathbb{Z}_3$. In dimension 3, the paragroup with graph $\mathbb{A}_5$ corresponds to the quotient S_3/S_2 of symmetric grouups. The coupling systems are given by the duality coupling between thes paragroups and their duals. For the hyperfinite II_1

factor R the classification theorem yields the following results.

COROLLARY

All subfactors of index 2 of R are normal and are conjugate to $R^{\mathbb{Z}_2} \subset R$ and to $R \subset R \rtimes \mathbb{Z}_2$ for a free action of $\mathbb{Z}_2$ on R.

□

The normality is Goldman's theorem and the rest is Connes' classification of actions of $\mathbb{Z}_2$.

COROLLARY

All normal subfactors of index 3 of R are conjugate to $R^{\mathbb{Z}_3} \subset R$ and to $R \subset R \rtimes \mathbb{Z}_3$ or for a free action of $\mathbb{Z}_3$ on R.

All nonnormal subfactors of index 3 of R are conjugate to $R^{S_3} \subset R^{S_2}$, to $R \rtimes S_2 \subset R \rtimes S_3$ and to the Jones subfactor of index 3.

□

The first part of the corollary is Sutherland's description of normal subfactors as cross products together with Connes' classification of the actions of $\mathbb{Z}_3$. The second part of the corollary is new.

REMARKS

In each graph the distinguished vertices * is a vertex with the smallest entry of the Perron eigenvector μ.

In dimension less than 4 the maximum number of nonisomorphic coupling systems (and thus of nonisomorphic subfactors of same index < 4 of the hyperfinite II_1 subfactor, cf. the classification theorem) is 4 in dimension $4\cos^2\frac{\pi}{30}$.

REMARKS ON THE PROOF

The fact that the graphs must be Coxeter - Dynkin $\mathbb{A}$, $\mathbb{D}$, $\mathbb{E}$ diagrams follows from Kronecker's theorem on matrices of norm < 2.

The position of the star at a vertex with minimum Perron eigenvector entry is a corollary of the axioms.

The contragredient map is determined by the restrictions on the number of paths joining vertices imposed by the biunitarity of the connection.

The exhaustion of the equivalence classes of connections proceeds as follows. Along each leg of the diagrams there is enough freedom given by the equivalence to make all choices equivalent; this takes care of the case $\mathbb{A}_n$. In the cases $\mathbb{D}_n$ and $\mathbb{E}_n$ there remains a choice at the triple point of the diagram which yields nonequivalent connections.

Although there are no subfactors with diagrams $\mathbb{D}_{2k+1}$ or $\mathbb{E}_7$, there are connections on each of these, as well as on $\mathbb{D}_{4k}$ and respectively $\mathbb{D}_{4k+2}$ with the endpoints of the short legs selfcontragredient and respectively contragredient to each other. The connections are given by the same formulae as the ones in the theorem, but fail the global axioms. The graphs $\mathbb{D}_{2k+1}$ and $\mathbb{E}_7$ are eliminated by the parallel transport axiom, while $\mathbb{D}_{2k}$ with the wrong contragredient map satisfy the parallel transport axiom but fail the global contragredient axiom. These connections can be used to construct string model subfactors with correct index. When the parallel transport axiom is not satisfied, the induction - restriction graphs might not be the starting ones. If the parallel transport axiom is satified but the global contragredient is not, the induction - restriction graphs will be the initial ones, but the contragredient map will be different from the map started with.

BIBLIOGRAPHICAL NOTES

There are several lines of research brought together by the approach described in the sequel. The two main directions are the study of subfactors and the study of actions of groups (or group - like objects) on algebras. A first result common to both was Goldman's theorem [Go,1960] stating that a subfactor of index 2 is the fixed point algebra of a twisted action of $\mathbb{Z}_2$. Together with the classification by Connes [Co2,1977] of the actions of $\mathbb{Z}_n$, this result shows that up to conjugacy there is only one subfactor of index 2 of the hyperfinite II_1 factor.

In a breakthrough paper [Jo3,1982] Vaughan Jones showed that the values less than 4 of the index are $4\cos^2(\pi/n)$. He did the first systematic study of the basic construction, which had appeared in the thesis of Skau [Sk,1977] and iterated it to obtain the tower of extensions associated to $N \subset M$. The projections given by conditional expectations satisfied the Jones relations $e_i e_{i\pm1} e_i = \tau e_i$ and $e_i e_j = e_j e_i$ for $|i - j| \geqq 2$, which gave representations of the braid group yielding new polynomial invariants for knots and links [Jo5,1985]. These relations also appear in statistical mechanics as the Temperley - Lieb relations [TL,1971]. Jones showed that conversely, the subalgebra generated by $e_1, e_2, \ldots$ inside the algebra generated by such projections $e_0, e_1, e_2, \ldots$ for $\tau = (4\cos^2(\pi/n))^{-1}$ gives examples of subfactors with index $4\cos^2(\pi/n)$ (the Jones factors, having graphs the Coxeter - Dynkin diagrams A_{n-1}.) The graph of the tower of relative commutants, which was introduced by Jones as a conjugacy invariant of subfactors, was shown to have floors stabilizing eventually to a Coxeter - Dynkin A, D or E diagram which repeats itself (since the vertices of the floors match); this raised the question of describing this graph up to the stabilizing level and proving an analogous result for arbitrary finite index. Techniques in this direction were developed independently by Wenzl [We2,1985] and us [Oc3,1985]. The answer is provided by the introduction of the principal graph in [Oc3,1985] which was later shown to coincide with the induction - restriction graph described in the present paper; the diagrams of the tower of relative commutants are then path algebras on the principal graph. A path algebra of a graph was subsequently also introduced by Sunder [Su,1986]. The thesis of Wenzl contains general methods for finding the index and the relative commutant of subfactors given in a nonstandard form, with which he obtained the values $\sin^2(k\,\pi)/\sin^2((k+1)\,\pi)$ for the index of the subfactors coming from the Hecke algebras. He later computed the graph associated to these subfactors, also described independently by us. A. Wassermann [Wa,1987] has introduced a different method of computing the tower of relative commutants for quotients of groups. Using McKay's results on discrete subgroups of SU(2), Jones constructed examples of subfactors with graphs extended Coxeter - Dynkin diagrams $A^{\sim}$, $D^{\sim}$, $E^{\sim}$ and index 4. Various topics related to subfactors are discussed in the monography [GHJ].

These papers raised several questions about subfactors of index < 4 of the hyperfinite II_1 factor: whether there exist up to conjugacy only finitely many subfactors with the same index, whether the subfactor of index φ^2 (φ = Golden Ratio) is unique, whether the only subfactors of index 3 correspond to $\mathbb{Z}_3$ and S_3/S_2 and whether there are any nontrivial sequences in the subfactor which are centralizing the factor, answered in this paper.

For a subfactor with finite index $N \subset M$ Pimsner and Popa [PP1,1985] constructed an orthogonal basis of the module $_N M$, gave a general definition of the index and defined the entropy. They proved the gluing lemma for iterated basic constructions [PP2,1985] also obtained independently by us [Oc3,1985]. Several of these techniques and results were carried over to the infinite index case. The standard projections, the Jones relations and the module basis have been extended to subfactors of infinite index, (in the discrete - compact case) by Herman and us [HO] using the operator valued weights of Haagerup. Sorin Popa has studied infinite index subfactors which are amenable or have Kazhdan's property T [Po,1986]. The extension of the index to subfactors of arbitrary type was done by Kosaki [Ko,1986]. Phan Loi developed the type III theory further and showed among others that subfactors of finite index ι of a factor of type III_λ must be of type $III_{\lambda'}$, with $\lambda' = \lambda^{p/q}$ where $p,q \in \mathbb{N}$ and $p\,q \leqq \iota$ [Lo,1987].

A different direction of research, which was our point of departure, was the study of outer actions of groups on operator algebras. Alain Connes classified the actions of $\mathbb{Z}_n$ and $\mathbb{Z}$ on hyperfinite factors [Co1] [Co2] The former result was extended to actions of finite groups on the hyperfinite II_1 factor by V. Jones [Jo1] while the latter was generalized to actions of discrete amenable groups on factors of type II, and more generally to actions of discrete amenable groups with trivial modular invariant on arbitrary factors by us [Oc1]. In unpublished work, we

classified prime actions of compact groups on type II factors, using the Takesaki duality for actions. The dual of a compact nonabelian group can be viewed as a quantum group (Hopf algebra, Kac algebra) which forced us to develop the ergodic theory for actions of these group - like objects. Our point of departure in 1985 was the remark that the golden ratio subfactor $N \subset M$ is the fixed point algebra of a homomorphism $\sigma: M \longrightarrow Mat_2 \otimes M$ which satisfies, up to an inner perturbation, the relation $\sigma \circ \sigma = \sigma \oplus id$, and thus $\{id, \sigma\}$ behaves like a group (or rather like a quantum group or Kac algebra.) Another remark was that the tower of extensions for a crossed product consists of the successive crossed products by the group and its dual. This proved to be the case in general, and the group like objects called paragroups were announced in [Oc2,1985] and presented at the AMS meeting in January 1986, together with a short proof of the classification of finite group actions using subfactors. This method of classification was extended to subfactors of finite index and finite depth and presented at meetings in July 1986.
The canonical shift unified such objects as the group multiplication, the comultiplication in Hopf algebras (for the case of crossed products) and the canonical endomorphism previously introduced by Doplicher and Roberts [DR] (for the case of embeddings corresponding to representations of compact groups.) The harmonic analysis on quotients of groups by nonnormal subgroups displayed a local structure, with all the information contained in numbers associated to cells. Such an approach could be obtained for general subfactors, with the graphs and cells coming from the induction and restriction of bimodules. The methods systematized the previously developed theory and explained the rigidity of the index and the apparition of the

diagrams A - D - E in the tower of relative commutants proved by V. Jones. The local approach was detailed enough to allow the description of all the subfactors of index less than 4, with the surprizing conclusion that the diagrams D_{odd} and E_7 do not appear as induction - restriction graphs of subfactors (since they are not geometrically flat.) These results were first obtained by computer modelling, and the computations involved in them proved to be very similar to those in the extensions by Pasquier of the Andrews - Baxter - Forrester models in statistical mechanics [Pa,1986].

REFERENCES

[Br] O. Bratelli: Inductive limits of finite dimensional C*-algebras. Trans. A.M.S. 171 (1972), 195-234.

[Co1] A. Connes: Une classification des facteurs de type III. Ann.Sci.Ec.Norm.Sup. 6 (1973) 133-252.

[Co2] A. Connes: Periodic automorphisms of the hyperfinite factor of type II_1. Acta Sci. Math. 39 (1977), 39-66.

[Co3] A. Connes: Outer conjugacy classes of automorphisms of factors. Ann. Sci. Ec. Norm. Sup. 8 (1975), 383 - 419.

[DR] S. Doplicher and J. Roberts: Endomorphisms of C*-algebras, cross products and duality for compact groups. Preprint.

[Go] M. Goldman: On subfactors of factors of type II_1. Mich. Math. J. (1960), 167-172.

[GHJ] F.Goodman, P. de la Harpe and V.F.R. Jones: Coxeter - Dynkin diagrams and towers of algebras I - III. Preprint.

[HO] R.Herman and A. Ocneanu: in preparation.

[Ho] A. Hoffman: On limit points of spectral radii of nonnegative symmetric integral matrices. Springer Lecture Notes 303 (1972).

[Jo1] V.F.R. Jones: Actions of finite groups on the hyperfinite II_1 factor. Mem. A.M.S. 237 (1980).

[Jo2] V.F.R. Jones: Sur la conjugaison des sous-facteurs des facteur de type II_1. C.R.Ac.Sci.Paris t. 286 (1977), 597-598.

[Jo3] V.F.R. Jones: Index for subfactors. Invent. Math., 71 (1983), 1-25.

[Jo4] V.F.R. Jones: Index for subrings of rings. Contemp. Math. 43 (1985), 181-190.

[Jo5] V.F.R. Jones: A polynomial invariant for knots via von Neumann algebras. Bull. A.M.S. 12 (1985), 103-112.

[Ko] H. Kosaki: Extension of Jones' theory on index to arbitrary factors. J. Funct. Analysis, 66 (1986), 123-140.

[Lo] P. Loi: On the theory of index and type III factors. To appear in Comptes Rendus Ac.Sci.Paris.

[MN] F. Murray and J. von Neumann: On rings of operators III - IV. Ann. Math. 37 (1936), 116-229; Ann. Math. 44 (1943), 716-808.

[Oc1] A. Ocneanu: Actions of discrete amenable groups on von Neumann algebras. Springer Lecture Notes 1138 (1980).

[Oc2] A. Ocneanu: Subalgebras are canonically fixed point algebras. A.M.S. Abstracts 6 (1986) 822-99-165.

[Oc3] A. Ocneanu: A Galois theory for von Neumann algebras. Notes of a UCLA lecture. Nov. 1985.

[Oc4] A. Ocneanu: Path algebras. Notes from a series of lectures,

A.M.S. meeting in Santa Cruz, 1986.

[Pa] V. Pasquier: Operator content of the A-D-E lattice models. CEN - Saclay preprint.

[PP1] M. Pimsner and S. Popa: Entropy and index for subfactors. Ann. Sci. Ec. Norm. Sup., t. 19 (1986), 57-106.

[PP2] M. Pimsner and S. Popa: Iterating the basic construction. INCREST preprint, 1986.

[Po] S. Popa: Correspondences. INCREST preprint, 1986.

[Sk] C. Skau: Finite subalgebras of a von Neumann algebra. J. Funct. Anal. 25 (1977), 211-235.

[Su] V. Sunder: A model for AF algebras and a representation of the Jones projections. Preprint 1986.

[TL] H.N.V. Temperly and E. Lieb: Relations between the percolation and colouring problems .... Proc. Roy. Soc. London A 322 (1971).

[Sh] J. Shearer: On the distribution of the maximum eigenvalue of graphs. I.B.M. preprint 1987.

[Wa] A. Wassermann: Ergodic actions of compact groups on operator algebras I - III. U.C.Berkeley preprints.

[We1] H. Wenzl: On sequences of projections. Preprint, Univ. of Pennsylvania, 1984.

[We2] H. Wenzl: Representations of the Hecke algebra and subfactors. Thesis, Univ. of Pennsylvania, 1985.

On Amenability in Type II_1 Factors

Sorin Popa

1. **Introduction**. In [8] we gave a proof of Connes' fundamental theorem that injectivity implies hyperfiniteness for finite von Neumann algebras. It consisted in first proving a certain Følner type condition for injective algebras, and then in combining this Følner condition with a local Rohlin lemma to get a local approximation property. Using a maximality argument, the latter gives the proof. We show here that by using a slightly stronger local Rohlin lemma, one can derive a similar proof of injectivity implies hyperfiniteness from Connes' Følner type condition ([1],[2]). Although we will use the same ideas as in [8], by reordering some of the steps in [8] the proof will become here more conceptual and shorter. In fact the proof in [8] can also be shortened using this reordering of the arguments.

1.1 **Notation**. Throughout this paper, M will denote a finite von Neumann algebra with a fixed normal finite faithful trace τ ($\tau(1) = 1$), $\|x\|_2 = \tau(x^*x)^{1/2}$, $x \in M$ will be the Hilbert norm given by τ, and $L^2(M,\tau)$, the completion of M in this norm. We let $\mathcal{U}(M)$ denote the unitaries in M.

1.2 **Injectivity**. The algebra M is called injective if there exists an M-invariant state ψ_0 on $\mathcal{B}(L^2(M,\tau))$, i.e., ψ_0 satisfies $\psi_0(xT)=\psi_0(Tx)$ for all $x \in M$, $T \in \mathcal{B}(L^2(M,\tau))$. Such a state is called a hypertrace on M and may be regarded as the operator algebra analogue of an invariant mean on an amenable group (see [2]).

1.3 Hyperfiniteness. The algebra M is called hyperfinite or approximately finite dimensional, if there exists an increasing sequence of finte dimensional subalgebras M_n for which $(\cup M_n)^- = M$ (the bar indicating weak closure - this of course implicitly means that the predual of M is separable).

1.4. Murray-von Neumann global approximation property. It was proved in [6] that for von Neumann algebras M with separable predual, the hyperfinitenss (as defined in 1.2) is equivalent to the following approximation property: given any finite subset $F=\{v_1,\dots,v_n\}$ of unitary elements in M and $\varepsilon>0$, there exists a finite dimensional *-sualgebra $M_1 \subset M$ and elements $v_1',\dots,v_m' \in M_1$ so that $\|v_i - v_i'\|_2 < \varepsilon$ for all i. This property is in fact independent of the separability of M, and when speaking about hyperfiniteness of an arbitrary finite von Neumann algebra M, we will mean that each countable decomposable summand of M has this approximation property.

1.5. The local approximation property. It was pointed out, implicitly in [3] and explicitly in [8], that in fact for M to have the property 1.4, it is sufficient that any reduced algebra of M satisfies the following: given any finite set of unitary elements $v_1,\dots,v_n$ and $\varepsilon>0$, there exists a projection f and a matrix subalgebra M_0 having f as unity,so that $\|v_i f v_i^* - f\|_2 < \varepsilon \|f\|_2$, and so that M_0 contains elements $v_1',\dots,v_n'$ with $\|f v_i f - v_i'\| < \varepsilon \|f\|_2$, for all i. Indeed, by a simple maximality argument this local approximation property implies the global approximation property of Murray and von Neumann.

Thus to prove that an injective finite algebra is hyperfinite (in the sense of 1.4) we only need to prove the local approximation. Moreover, since the injectivity of M is clearly inherited by its reduced algebras, we only need to prove that if M is injective and $v_1,\dots,v_n \in M$, $\varepsilon>0$, then there exists a projection $f \in M$ and a matrix algebra M_0 with unity f, which contains some elements $v_1',\dots,v_n' \in M_0$ for which

$$\|f v_i f - v_i'\|_2 \le \varepsilon \|f\|_2 \text{ and } \|v_i f v_i^* - f\|_2 < \varepsilon \|f\|_2 \text{ for all i.}$$

1.6. A digression on the Connes-Feldman-Weiss theorem. Let $(X,\mathcal{X},\mu)$ be a nonatomic probability space and G a discrete countable group acting properly outer and measure preserving on X. Let A = $L^\infty(X,\mu)$, $M=A\rtimes G$. The Connes-Feldman-Weiss theorem [3] states that given any finite set $F\subseteq G$ and any $\varepsilon>0$, there exists a matrix *-subalgebra $M_1\subset M$ with matrix unit $\{e_{ij}\}_{i,j}$ so that the diagonal $A_1=\{e_{ii}\}_i''$ is in A, the normalizer of A_1 in M_1 is included in the normalizer of A in M, and there exist $v_1',\ldots,v_n'$ in the normalizer of A_1 with $\|v_i - v_i'\|<\varepsilon$ for all i.

The key part in the proof of this theorem consists in proving a suitable local approximation property similar to 1.5. In its simplest form the proof of the local approximation goes as follows: Let $\{u_g\}_{g\in G}\subset M$ be the unitaries implementing the action of G on A. Let $F'\subset G$ be a finite set and $\varepsilon>0$.By what we call the commutative local Rohlin lemma, there is a nonzero projection $e\in A$ so that $eg(e)=0$ if $g\in F'-\{1\}$, or equivalently, $eu_ge = \tau(u_g)e$. In particular, let $F'=K^{-1}FK$ where $1\in F$, $K\subset G$ are finite sets. If we denote $e_{g'g} = u_{g'}eu_g{}^{-1}$, then $\{e_{g'g}\}_{g',g\in K}$ gives a matrix unit for a matrix subalgebra M_0 with support

$$f = \sum_{g\in K} u_geu_g{}^{-1}$$

and by the local Rohlin lemma for each $h\in F$ we have

$$fu_hf = \sum_{g,g'\in K} \tau(u_{g'} {}_{hg})e_{g'g} \in M_0.$$

Now by the Følner condition of G, if F is fixed, then K may be chosen so that $\mathrm{card}(FK\setminus K) < \varepsilon\,\mathrm{card}\ K$, which in turn means that

$$\|u_hfu_h{}^{-1} - f\|_2{}^2 < \varepsilon\|f\|_2{}^2,\ h\in F.$$

This shows that $\{u_h\}$ can be approximated under an almost invariant projection by a matrix algebra.

The reader should keep in mind this argument because it is an approximate version of it that we will prove in the next section.

2. Proof of the local approximation property. In this section we shall, for simplicity, assume that M is a separable type II_1 factor.

2.1 The noncommutative local Rohlin lemma. The local Rohlin lemma we need here is slightly stronger than the one in [7], [8]. Its proof is a simple consequence of results in [8] as will be shown in

Section 3.

2.1.1. **Lemma**. Let $F' \subset M$ be a finite set and $\varepsilon' > 0$. There exists a finite set $\mathcal{P}'$ of mutually orthogonal projections e' in M so that $\tau(\Sigma e') \geq 1-\varepsilon'$, and so that for each $e' \in \mathcal{P}'$,

$$\sum_{y \in F'} \| e' y e_i' - \tau(y) e' \|_2^2 < \varepsilon' \| e' \|_2^2.$$

Now let $1 \in F \subset M$ be a finite set of unitary elements in M. Let $K = \{x_1, \ldots, x_m\} \subset M$ be a finite set with $\tau(x_j^* x_i) = \delta_{ij}$. Apply 2.1.1 to

$$F' = \{x_j^* v x_i \mid 1 \leq i,j \leq m,\ v \in F\},$$

$\varepsilon' > 0$, and let $e' \in \mathcal{P}'$. Then we have

$$\sum_{y \in F'} \| e' x_j^* v x_i e' - \tau(x_j^* v x_i) e' \|_2^2 < \varepsilon' \| e' \|_2^2.$$

In particular, since $1 \in F$, we have:

$$\| e' x_j^* x_i e' - \delta_{ij} e' \|_2^2 < \varepsilon' \| e' \|_2^2.$$

Thus when restricted to the projections e', x_j behave almost like partial isometries with mutually orthogonal left supports. A routine deformation argument (see, e.g. [4]) then implies the existence of a set of partial isometries $u_1, \ldots, u_m \in M$ with :

(1) $u_j^* u_i = \delta_{ij} e$, where $e \leq e'$ is a projection with $\| e' - e \|_2 < \delta(\varepsilon') \| e \|_2$;

(2) $\| u_i - x_i e \|_2 < \delta(\varepsilon') \| e \|_2$, $1 \leq i \leq m$;

(3) $\sum_{v \in F} \| u_j^* v u_i - \tau(x_j^* v x_i) e \|_2^2 < \delta(\varepsilon') \| e \|_2^2$, $1 \leq i,j \leq m$,

where $\delta(\varepsilon') \to 0$ when $\varepsilon' \to 0$.

It follows that if $e_{ij} = u_i u_j^*$ and $f = \Sigma e_{ii}$ then $\{e_{ij}\}_{1 \leq i,j \leq m}$ is a matrix unit for a matrix subalgebra $M_0 \subset M$ with support f and by (3) we have

$$(4)\ \sum_{v \in F} \| f v f - \sum \tau(x_j^* v x_i) e_{ji} \|_2^2 = \sum_{i,j} \sum_{v \in F} \| e_{jj} v e_{ii} - \tau(x_j^* v x_i) e_{ji} \|_2^2$$
$$= \sum_{i,j} \sum_{v \in F} \| u_j^* v u_i - \tau(x_j^* v x_i) e \|_2^2$$
$$< m^2 \delta(\varepsilon') \| e \|_2^2 = m \delta(\varepsilon') \| f \|_2^2 = \delta_1(\varepsilon') \| f \|_2^2.$$

Moreover, by (1), if we denote by $\mathcal{P}$ the set of all of the above projections e ($\leq e' \in \mathcal{P}'$), then the e's are mutually orthogonal and

$$\tau(\Sigma_{e \in \mathcal{P}}\, e) \geq 1 - \delta_2(\varepsilon').$$

Let us summarize all of these properties:

2.1.2 **Lemma**. Let $F \subset \mathcal{U}(M)$ be a finite set. Let $K = \{x_1, \ldots, x_m\} \subset M$ with $\tau(x_j^* x_i) = \delta_{ij}$. Let also $\varepsilon'' > 0$. Then there exists a finite set $\mathcal{P}$ of mutually orthogonal projections e in M with the properties

(i) $\tau(\Sigma e) \geq 1 - \varepsilon''$;

(ii) for each e there exist partial isometries $u_1,\ldots,u_m$ in M so that

1°. $u_j^*u_i = \delta_{ij}$;

2°. $\|u_i - x_i e\|_2 < \varepsilon'' \|e\|_2$, $1 \le i \le m$;

3°. $\|u_j^* v u_i - \tau(x_j^* v x_i) e\|_2 < \varepsilon'' \|e\|_2$, for all $1 \le i,j \le m$ $v \in F$;

4°. if $e_{ij} = u_i u_j^*$ is the matrix unit associated to $\{u_i\}$ and $f = \Sigma_i e_{ii}$, then for each $v \in F$ we have:

$$\sum_{v\in F} \|fvf - \sum_{i,j} \tau(x_j^* v x_i) e_{ji}\|_2^2 \le \varepsilon'' \|f\|_2^2.$$

2.2 Connes' Følner type condition. Given a Hilbert space H, we may associate with each $\xi \in H$ a linear functional ξ^*in the dual Hilbert spaceH^* by letting $\xi^*(\eta) = \eta \cdot \xi$. Let $L^2(M,\tau) \bar{\otimes} L^2(M,\tau)^*$ be identified with the set of Hilbert Schmidt operators on $L^2(M,\tau)$ on which M acts by left and right multiplication. The norm and the corresponding dot product on $L^2(M,\tau) \bar{\otimes} L^2(M,\tau)^*$ are denoted by $\|\ \|_{HS}$ and $\langle,\rangle_{HS}$, respectively.

2.2.1. **Theorem** ([1]). Suppose that M is injective. Let $F \subset \mathcal{U}(M)$ be a finite set and $\varepsilon > 0$. There exists a finite dimensional projection Q on $L^2(M,\tau)$ so that

$$\sum_{v\in F} \|vQv^* - Q\|_{HS}^2 < \varepsilon^2 \|Q\|_{HS}^2.$$

But a finite dimensional projection Q on $L^2(M,\tau)$ is of the form

$$\Sigma_{i=1}^{m} \xi_i \otimes \xi_i^*,$$

where ξ_i are mutually orthogonal vectors of norm one in $L^2(M,\tau)$. Since M is a dense vector subspace of $L^2(M,\tau)$, by applying the orthonormalization algorithm, it follows that the orthonormal system $\xi_1,\ldots,\xi_m$ can be approximated arbitrarily well by an orthonormal system $x_1,\ldots,x_m$ in M. Thus in 2.2.1 we can assume that $Q = \Sigma x_i \otimes x_i^*$, where $x_i \in M$, $\tau(x_j^* x_i) = \delta_{ij}$.

2.3. Combining the local Rohlin lemma with the Følner condition. For the next computations it is important to note that if $p,q \in N$ are mutually orthogonal projections and $x,x',y,y' \in M$, then the vectors $x'p \otimes x^*$, $y'q \otimes y^*$ are mutually orthogonal in $L^2(M,\tau) \bar{\otimes} L^2(M,\tau)^*$. Indeed we have

$$\langle x'p \otimes x^*, y'q \otimes y^* \rangle_{HS} = \tau(qy'^*x'p)\tau(x^*y) = \tau(y'^*x'pq)\tau(x^*y) = 0.$$

Now assume that the separable type II_1 factor M is injective.

Let $F \subset \mathcal{U}(M)$ be a finite set and let $\varepsilon > 0$. By 2.2 there is a set $K = \{x_1, \ldots, x_m\} \subset M$ with $\tau(x_j^* x_i) = \delta_{ij}$ so that $Q = \Sigma_i x \otimes x_i^*$ satisfies

$$\sum_{v \in F} \| vQv^* - Q \|_{HS}^2 < (\varepsilon/8)^2 \| Q \|_{HS}^2.$$

Let $\varepsilon'' > 0$ be chosen so that $\varepsilon'' \le \varepsilon/12m^{1/2}$ and so that if p is a projection in M with $\tau(p) \ge 1 - \varepsilon''$, then $\tau(x_i^* x_i p) \ge 1/2$, for all i. Apply the local Rohlin lemma 2.1.2 to the finite sets F,K, and to $\varepsilon'' > 0$. With the notations in 2.1.2 we get from the definition of the Hilbert-Schmidt norm and from the above observation:

$$\sum_{e \in \mathcal{P}} \sum_{v \in F} \| v(\sum_i x_i e \otimes x_i^*)v^* - \sum_i x_i e \otimes x_i^* \|_{HS}^2 =$$

$$= \sum_{v \in F} \| v(\Sigma_i x_i (\Sigma e) \otimes x_i^*)v^* - \Sigma_i x_i (\Sigma e) \otimes x_i^* \|_{HS}^2$$

$$\le \sum_{v \in F} \| v(\Sigma_i x_i \otimes x_i^*)v^* - \Sigma_i x_i \otimes x_i^* \|_{HS}^2$$

$$< (\varepsilon/8)^2 \| \Sigma_i x_i \otimes x_i^* \|_{HS}^2$$

$$\le 2\,(\varepsilon/8)^2 \| \Sigma x_i \,(\Sigma e) \otimes x_i^* \|_{HS}^2$$

$$= 2\,(\varepsilon/8)^2 \sum_{e \in \mathcal{P}} \Sigma_i \| x_i e \otimes x_i^* \|_{HS}^2.$$

It follows that there exists an $e \in \mathcal{P}$ so that

$$\| v(\Sigma x_i e \otimes x_i^*)v^* - \Sigma x_i e \otimes x_i^* \|_{HS} < \varepsilon/4 \; \| \Sigma x_i e \otimes x_i^* \|_{HS},$$

for all $v \in F$.

If u_i are the partial isometries approximating $x_i e$ as in 2.2, then we get

$$\| x_i e \otimes x_i^* - u_i \otimes x_i^* \|_{HS} = \| x_i e - u_i \|_2 \le \varepsilon'' \| e \|_2 = \varepsilon'' \; \| u_i \otimes x_i^* \|_{HS},$$

and thus

(*)
$$\begin{aligned} \| v(\Sigma_i u_i \otimes x_i^*)v^* - \Sigma_i u_i \otimes x_i^* \|_{HS} &< \varepsilon/4 \; \| \Sigma_i u_i \otimes x_i^* \|_{HS} + 3m \| x_i e - u_i \|_2 \\ &\le (\varepsilon/4 + 3m^{1/2} \varepsilon'') \| \Sigma_i u_i \otimes x_i^* \|_{HS} \\ &\le (\varepsilon/2) \| \Sigma_i u_i \otimes x_i^* \|_{HS}. \end{aligned}$$

Recall that with the notations $e_{ij} = u_i u_j^*$, $f = \Sigma e_i = \Sigma u_i^* u_i$ and $\lambda_{ji} = \tau(x_j^* v x_i)$, we have by 2.1.2:

(i).
$$\| fvf - \sum_{i,j} \lambda_{ji} e_{ji} \|_2 < \varepsilon'' \| f \|_2.$$

Moreover, since $\| v(\Sigma_i u_i \otimes x_i^*)v^* \|_{HS} = \| \Sigma_i u_i \otimes x_i^* \|_{HS} = \| f \|_2$, we also have:

(**)
$$\begin{aligned} \| v(\Sigma_i u_i \otimes x_i^*)v^* - \Sigma_i u_i \otimes x_i^* \|_{HS}^2 &= \\ &= 2 \| f \|_2^2 - 2\mathrm{Re} \sum_{i,j} \tau(u_j^* v u_i) \tau(x_i^* v^* x_j) \\ &= 2 \| f \|_2^2 - 2\mathrm{Re}\; \tau(\sum_{i,j} \overline{\lambda}_{ji} e_{ij} v) \end{aligned}$$

$$\geq 2\|f\|_2^2 - 2\mathrm{Re}\ \tau(fv^*fv) - 2\varepsilon'' \|f\|_2^2.$$
$$= \|vfv^* - f\|_2^2 - 2\varepsilon''\|f\|_2^2,$$

Thus by (*) and (**) we get:

(ii). $\|vfv^* - f\|_2^2 < \varepsilon\|f\|_2^2.$

But (i) and (ii) show that M has the local approximation property. Thus M is hyperfinite.

3. Some Comments. As one can see, if one assumes the local Rohlin lemma and the Følner type condition, then the proof **2.3** of the local approximation property (and thus, by **1.5**, of the whole thoerem) is very much the same as in the commutative ergodic case 1.6, and is therefore quite simple. In fact, if we take $M=\lambda(G)''$ to be the von Neumann algebra of the left regular representation of an amenable group G, then **2.3** gives the same approximation for a finite set of elements in G as does **1.6**. This observation gives a rough idea of how one can "chop up" the elements of G to build the matrix algebras approximating $\lambda(G)''$.

On the other hand, both the Følner type result and the local Rohlin lemma have rather simple and elementary proofs:

3.1 The Connes' Følner type condition. This is fully dicussed in [2], where it is pointed out that there is a perfect analogy between its proof and the classical proof of Namioka (using Day's trick) to the Følner theorem. Indeed, the rôle of the invariant mean on G is played here by an M-invariant state on $\mathcal{B}(L^2(M,\tau))$, the hypertrace. Using the functional analysis trick of Day, one gets from this M-invariant (non)normal state normal states on $\mathcal{B}(L^2(M,\tau))$ almost invariant to finite subsets of M. But this means we have an almost invariant vector in $\mathcal{B}(L^2(M,\tau))_* = c_1(L^2(M,\tau))$, or by taking square roots, in

$$c_2(L^2(M,\tau)) = L^2(M,\tau)\bar{\otimes}L^2(M,\tau)^*.$$

Then by suitable manipulation of the spectral theorem, one gets an almost invariant projection in $L^2(M,\tau)\bar{\otimes}L^2(M,\tau)^*$.

3.2. The proof of the local Rohlin lemma 2.1. This can be easily derived from results in [7]. In turn, those results have elementary proofs. We will show here how one can deduce a slightly more general result than 2.1.1, namely:

3.2.1 Lemma (see 2.3.3 in [9]). Let M be a separable type II_1 factor and $N \subset M$ a subfactor with $N' \cap M = \mathbb{C}$. Let F' be a finite set and $\varepsilon > 0$. Then there exists a partition of unity $e_1, \ldots, e_n \in N$ with projections in M so that

$$\sum_{y \in F'} \| \sum_i e_i y e_i - \tau(y) 1 \|_2^2 < \varepsilon^2.$$

Proof: By [7] there is a hyperfinite II_1 subfactor $R \subset N$ so that $R' \cap M = \mathbb{C}$. In a hyperfinite II_1 factor there exists a decreasing sequence of subfactors R_n with $\cap R_n = \mathbb{C}$, and with $R_n' \cap R = \mathbb{C}$ for each n. To see this, let the group S_∞ of all the permutations over $\mathbb{N}$ act on R properly outer and ergodically (e.g., as Bernoulli shifts), and let $K_n \subset S_\infty$ be an increasing sequence of finite subgroups with $\cup K_n = S_\infty$. If $R_n = R^{K_n}$ are the algebras fixed by K_n, then R_n will do. For $B \subset M$ a von Neumann subalgebra we let E_B denote the unique normal trace preserving conditional expectation onto B. Then by 1.2 in [7] we have $\| E_{R_n}(x) - \tau(x) \|_2 \to 0$ for each $x \in M$. Let n be such that

$$\sum_y \| E_{R_n}(y) - \tau(y) \|_2^2 < (\varepsilon/4)^2.$$

Since $R_n' \cap R = \mathbb{C}$, by [7] there exists a finite dimensional abelian subalgebra $A_0 \subset R_n$ so that

$$\sum_y \| E_{A_0' \cap R}(E_R(y)) - \tau(y) 1 \|_2^2 < (\varepsilon/4)^2.$$

Moreover, by [7] there exists a finite refinement A_1 of A_0 in R so that if $y' = y - E_R(y)$, then

$$\sum_y \| E_{A_1' \cap M}(y') \|_2^2 < (\varepsilon/2)^2.$$

We thus get

$$\begin{aligned} \sum_y \| E_{A_1' \cap M}(y) - \tau(y) 1 \|_2^2 &\le 2 \sum \| E_{A_1' \cap M}(y') \|_2^2 + \\ &+ 2 \sum \| E_{A_1' \cap M}(E_R(y)) - \tau(y) 1 \|_2^2 \\ &\le \varepsilon^2. \end{aligned}$$

Thus, if $e_1, \ldots e_n$ are the minimal projections of A_1 then we are done. QED

3.3. General finite von Neumann algebras. In order to make the proof of injectivity implies hyperfiniteness in §2 work for arbitrary finite von Neumann algebras, we only need a suitable more general Rohlin lemma:

3.3.1 General noncommutatve local Rohlin lemma. Let M be an arbitrary finite von Neumann algebra, and let N be a von Neumann subalgebra of M with $N' \cap M \subset N$, and let E be the canonical conditional expectation of N onto $\mathcal{Z}(N)$. Let also τ be a normal finite trace on M. Given any finite set $F' \subseteq M$ and $\varepsilon > 0$, there exists a partition of unity $e_1, \ldots, e_n$ with projections in N so that

$$\sum_y \| \sum_i e_i y e_i - E(y) \|_2^2 < \varepsilon^2.$$

3.4. The proof of injectivity implies hyperfiniteness in [8] uses a weaker local Rohlin lemma, which shows that given any finite set $F' \subset M$ one can fill up the identity with projections e in a given maximal abelian algebra $A \subset M$ so that eye are almost scalar multiples of e, for each $y \in F'$ (but the scalars are not necessarily of the form $\tau(y)$). In turn, however, to carry out the proof one then need a stronger Følner type condition: if M_1 is the extension of an injective algebra M by a maximal abelian subalgebra $A \subset M$ (in the sense of [5]) and $F \subset \mathcal{U}(M)$ is a finite set, then there exists an almost F-invariant projection in $M_1 \cap L^2(M_1)$. Then the proof proceeds as in §2.

3.5 The maximality argument that is used to prove that the local approximation property 1.5 implies the global approximation property 1.4 is given in detail in §3 of [8]. The idea of the proof is as follows:

If 1.5 holds true, then given any finite set $y_1, \ldots, y_n \in M$, there is a finite dimensional *-subalgebra $M_0 \subset M$, or more generally, a type I weakly closed *-subalgebra, with $1_{M_0} = f \in M$ satisfying $\| f y_i - y_i f \|_2 \leq (\varepsilon/3) \| f \|_2$, and $\| f y_i f - E_{M_0}(f y_i f) \|_2 < (\varepsilon/3) \| f \|_2$ for all i. But then

(*) $$\| (y_i - (1-f) y_i (1-f)) - E_{M_0}(f y_i f) \|_2^2 \leq \varepsilon^2 \| f \|_2^2$$

for $1 \leq i \leq n$.

Moreover we may take M_0 to be maximal, under the inclusion order, among all other type I subalgebras satisfying (*). Intuitively, the elements involved in (*) can be illustrated like in Figure 1 below, where the outermost shaded region represents $y - (1-f) y (1-f)$, and the element $E_{M_0}(fyf)$ is somewhere in the small top left square.

Now if (*) is not global, that is if $f \neq 1_M$, then we may apply 1.5 again, this time to the finite set $(1-f) y_i (1-f)$, $1 \leq i \leq n$, to get a new type I *-subalgebra $M_1 \subset (1-f) M (1-f)$ with $1_{M_1} = f_1$ satisfying

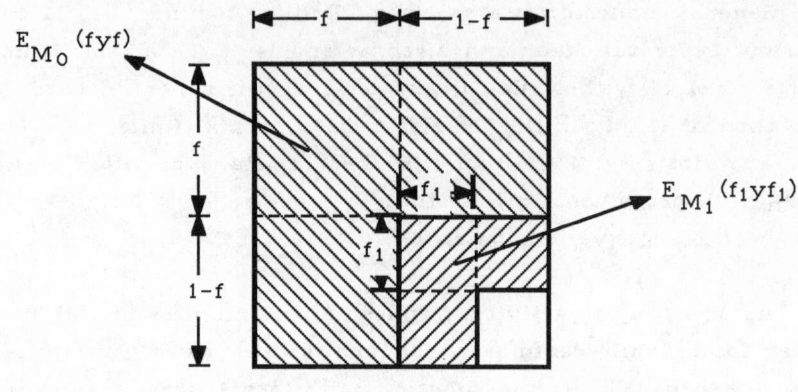

Figure 1.

$$(**)\qquad \|((1-f)y_i(1-f) - (1-f-f_1)y_i(1-f-f_1)) - E_{M_1}(f_1y_if_1)\|_2^2 \leq \varepsilon^2\|f_1\|_2^2$$

for $1\leq i\leq n$. These elements are also illustrated in Figure 1 by the intermediate shaded region. But as the picture shows it, the inequalities (*) and (**) involve mutually orthogonal vectors, so by the Pythogorean theorem, we may add them up to get:

$$\begin{aligned}\|y_i - (1-f-f_1)y_i(1-f-f_1) - E_{M_0\oplus M_1}((f+f_1)y_i(f+f_1))\|_2^2 \\ = \|(y_i - (1-f)y_i(1-f)) - E_{M_0}(fy_if)\|_2^2 \\ + \|((1-f)y_i(1-f) - (1-f-f_1)y_i(1-f-f_1)) - E_{M_1}(f_1y_if_1)\|_2^2 \\ \leq \varepsilon^2\|f\|_2^2 + \varepsilon^2\|f_1\|_2^2 = \varepsilon\|f+f_1\|_2^2\end{aligned}$$

for $1\leq i\leq n$. Since $M_0\oplus M_1$ is also of type I, this contradicts the maximality of M_0. Thus $1_{M_0} = 1_M$, and since a type I finite algebra is clearly hyperfinite, 1.4 is proved.

Acknowledgement

This work was completed during my stay as a visiting professor at the mathematics department of the University of California Los Angeles. I wish to express my gratitude to Professors R. Blattner, E. Effros, and M. Takesaki for their warm hospitality, and continuous help. I would also like to thank Professor E. Effros for his suggestions in writing the final version of the paper.

References

1. Connes, A.: Classification of injective factors, Ann. Math. 104 (1976), 73-115.

2. Connes, A. On the classification of von Neumann algebras and their

automorphisms, Symposia Math., XX, 435-478, Academic Prees, 1978.

3. Connes, A.; Feldman, J.; Weiss, B.: An amenable equivalence relation is generated by a simple transfomation, Ergodic Theory and Dynamical Systems, 1 (1981), 431-450.

4. Dixmier, J.: Les algebres d'operateurs dans l'espace hilbertien, Gauthier-Villars, Paris 1957, 1969.

5. Jones, V.: Index for subfactors, Invent. Math. 72 (1983), 1-25.

6. Murray, F; von Neumann, J.: On rings of operators IV, Ann Math. 44 (1943), 716-808.

7. Popa, S.: On a problem of R. V. Kadison on maximal abelian *-subalgebras in factors, Invent. Math. 65 (1981), 269-281.

8. Popa, S.: A short proof of injectivity implies hyperfiniteness for finite von Neumann algebras, J. Operator Theory, 16 (1985)

9. Popa, S.: Correspondences, INCREST, preprint 1986.

An index for continuous semigroups of *-endomorphisms of $\mathfrak{B}(\mathfrak{H})$

by

Robert T. Powers*
Department of Mathematics
University of Pennsylvania
Philadelphia, Pa 19104

* Supported in part by a National Science Foundation Grant

In this talk I will be discussing continuous semigroups of *-endomorphisms of $\mathfrak{B}(\mathfrak{H})$. Let me explain the motivation for considering such semigroups. In the "old" formulation of quantum mechanics one has the following mathematical framework. States of the physical system are described by a wave function $\Psi(t)$ where t time and $\Psi(t)$ is a unit vector in a Hilbert space $\mathfrak{H}$. The dynamics the system is given by the Schrödinger equation $i\partial\Psi/\partial t = H\Psi$ where is a self adjoint operator call the Hamiltonian. As is well known from Stone's theorem the solution the the Schrödinger equation is given by a strongly continuous one parameter unitary group U(t) so $\Psi(t) = U(t)\Psi(0)$. The mathematical difficulties associated with th picture is that it is not always clear that the Hamiltonians one encounters are in fact self adjoint. The Hamiltonians one encount in physic text book are usually presented as differential operator For example the Hamiltonians for the harmonic oscillator and the Hydrogen atom are given as

$$H_1 = -\tfrac{1}{2}\frac{d^2}{dx^2} + \tfrac{1}{2}x^2 \qquad H_2 = -\Delta - \frac{e^2}{r}$$

where H_1 acts on $L^2(\mathbb{R})$ and H_2 acts on $L^2(\mathbb{R}^3)$. One easily checks t the differential operators one encounters in physics are hermitian (i.e. $(f,Hg) = (Hf,g)$ for all $f,g \in \mathfrak{D}(H)$) usually by using integration by parts. But to check for self adjointness is more difficult. Sometimes an hermitian operator may fail to be self adjoint but it is possible to extend the operator to a self adjoin operator. The well known theorem of von Neumann says that an hermitian operator H has a self adjoint extension to an operator H if and only if the deficiency spaces $\mathfrak{D}_+$ and $\mathfrak{D}_-$ have the same

mension where $\mathcal{D}_{\pm} = \{f \in \mathfrak{H}; (f,(H \pm i)g) = 0$ for all $g \in \mathcal{D}(H)\}$ or
quivalently, $\mathcal{D}_{\pm} = \{f \in \mathcal{D}(H^*); H^* = \pm if\}$. The important point is that
determine whether an hermitian operator has a self adjoint
tension one only has to compute to indices $n_+ = \dim(\mathcal{D}_+)$ and $n_- =$
$m(\mathcal{D}_-)$. If these numbers are equal you can extend the Hamiltonian H
a self adjoint operator and thereby define a dynamics for your
ysical system. For example the operator $H = -i\, x(d/dx)x$ acting on
$^{\infty}$-function with compact support in $L^2(-\infty,\infty)$ has deficiency indices
,1) and, thus, it has a self adjoint extension while the operator
$= -i/2(x^3(d/dx) + (d/dx)x^3)$ has deficiency indices (0,2) so it does
t have a self adjoint extension.

This is the situation for the "old" formulation of quantum
chanics. What about the "new" version of quantum mechanics? The
w version of quantum mechanics as expressed in 1964 by Haag and
stler in [4] says that there is a C^*-algebra $\mathfrak{A}$ the algebra of
servables. States of the physical system are given by states ω of
and the dynamics is given by one parameter group of $*$-automorphisms
. The generator of a group of $*$-automorphisms α_t is a $*$-derivation
We define $\delta(A) = \lim_{t \to 0^+} (\alpha_t(A) - A)/t$ for all $A \in \mathcal{D}(\delta)$ where
$\delta)$ is the set of all $A \in \mathfrak{A}$ where the limit exists in the
propriate topology (the uniform or norm topology if we are working
th a C^*-algebra and the σ-weak topology if we are working with a
n Neumann algebra). One checks that δ has the property that $\mathcal{D}(\delta)$
a dense (in the appropriate topology) $*$-subalgebra of $\mathfrak{A}$ and δ is a
osed $*$-derivation of $\mathcal{D}(\delta)$ into $\mathfrak{A}$ i.e.,

(i) if $A,B \in \mathcal{D}(\delta)$ then $AB \in \mathcal{D}(\delta)$ and $\delta(AB) = \delta(A)B + A\delta(B)$.

(ii) if $A \in \mathcal{D}(\delta)$ then $A^* \in \mathcal{D}(\delta)$ and $\delta(A^*) = \delta(A)^*$.

One question one of importance is given a closed densely defin
$*$-derivation when is it the generator of a one parameter group of $*$-automorphisms. In chapter 3 of Bratteli and Robinson's book [2] there are two theorem which tell you when a closed densely defined $*$-derivation is the generator of a group of $*$-automorphisms. One theorem gives the conditions for a C^*-algebra and the second theore gives the conditions of the case of a von Neumann algebra. Without spelling out all the conditions in detail we point out there is an analogy with the case of hermitian operators on Hilbert space. If is a densely defined $*$-derivation then the analog of being hermitia is the condition that δ and $-\delta$ be dissipative which can be expresse as the condition that $\|A + \lambda\delta(A)\| \geq \|A\|$ for all $A \in \mathfrak{D}(\delta)$ and λ real The analog of being self adjoint is that the range of the maps $A \longrightarrow A \pm \delta(A)$ be all of the algebra $\mathfrak{A}$. For δ to be a generator of a gro of $*$-automorphisms both these conditions must be satisfied.

In physical applications the derivation δ is at best only defined heuristically typically in the form $\delta(A) = [H,A]$ where H is formal sum of infinitely many bounded operators or finitely many unbounded operators. There is usually a problem in defining δ and one makes a guess as to how to define δ in a given problem one may find that δ fails to satisfy the analog of self adjointness, the condition that the range of the maps $A \longrightarrow A \pm \delta(A)$ be all of $\mathfrak{A}$. Th it would be extremely useful to have some way of knowing when such δ can be extended to a $*$-derivation δ_1 defined on a larger domain s that δ_1 generates a group of $*$-automorphisms. It would be extremel useful if there was an index theory for $*$-derivation like the index theory of hermitian operators on Hilbert space.

At first glance it appears that all one has to do to is define indices of a $*$-derivation as the codimensions of the ranges of the s $A \longrightarrow A \pm \delta(A)$. However, this proves unfruitful because looking examples shows that these codimensions are always zero or infinity. derivation property forces the codimensions to be infinite if y are not zero. About three years ago I became interest in this blem inspired by the work of V. Jones which developed an index ory for type II_1 factors. For the case of II_1 factors if one es the most naive definition of the index of a subfactor N of a e II_1 factor M one might define the index to be the codimension of n M viewed as a subspace of M. Again one would find the index is ays infinite. My thinking was if there is a way to associate a ite number to a subfactor of a II_1 factor there may be a way to ociate a finite index to a $*$-derivation. It turns out, so far, t the index I will talk about for $*$-derivations has not much to do h the index for subfactors of a type II_1 factor but it was the k of V. Jones that got me thinking about this problem.

Since we are at the beginning of a theory of index for erivations let us make our task as easy as we can. Therefore, I t to specialize to the easiest case I know of. First let us centrate semigroups of $*$-endomorphisms. One would expect ndomorphisms to be much easier to handle by analogy with the uation for Hilbert space. Suppose A is an hermitian operator and of the deficiency indices n_+ or n_- is zero. Then iA or -iA is generator of a strongly continuous one parameter semigroup of metries U(t) defined for $t \geq 0$. And for such a semigroup U(t) ing on a Hilbert space $\mathfrak{H}$ there is a decomposition $\mathfrak{H} = \mathfrak{M}_1 \oplus \mathfrak{M}_2$ of $\mathfrak{H}$ o the direct sum of orthogonal subspaces so that that U(t) and

$U(t)^*$ map both $\mathfrak{M}_1$ and $\mathfrak{M}_2$ into themselves and one has the decomposition $U(t) = V(t) \oplus S(t)$ with respect to these subspaces wh V(t) is unitary and S(t) is isomorphic to n copies (where $n = n_+$ or n_- which ever is not zero) of the shift operator $S_o(t)$ on $L^2(0,\infty)$ (where $(S_o(t)f)(x) = f(x - t)$ for $x \geq t$ and $(S_o(t)f)(x) = 0$ for $x <$ for all $f \in L^2(0,\infty)$). If $-d$ is the generator of $U(t)$ (so $U(t) = \exp(-td)$) then d has a decomposition $d = d_1 \oplus d_2$ on $\mathfrak{M}_1 \oplus \mathfrak{M}_2$ where d_1 skew adjoint (a self adjoint operator times *i*) and d_2 is isomorphic n copies of the differential operator d/dx on $L^2(0,\infty)$. We see that hermitian operators with one non zero deficiency index are not much more complicated than self adjoint operators. I do not know if the is a simple classification of all the hermitian operators with deficiency indices (1,1). In short, requiring one deficiency index to vanish makes life a lot simpler in the study of hermitian operators. The analogous requirement for *-derivations is to study *-derivations which generate a one parameter semigroup of *-endomorphisms.

We will focus our attention on continuous semigroups of *-endomorphisms of von Neumann algebras. These semigroups will be named and defined as follows.

Definition 1. We say $\{\alpha_t;\ t \geq 0\}$ is an E_o-semigroup of a von Neumann algebra M if the following conditions are satisfied.

i) α_t is a *-endomorphism of M for each $t \geq 0$.

ii) α_o is the identity endomorphism and $\alpha_t \circ \alpha_s = \alpha_{t+s}$ for all $t,s \geq 0$.

iii) For each $f \in M_*$ (the predual of M) and $A \in M$ the function $f(\alpha_t(A))$ is a continuous function of t.

The second simplification we will make is to restrict our attention to the case where M is $\mathfrak{B}(\mathfrak{H})$. The von Neumann algebra $\mathfrak{B}(\mathfrak{H})$ is the simplest von Neumann algebra that admits a proper *-endomorphisms.

An important example of an E_o-semigroup of $\mathfrak{B}(\mathfrak{H})$ is obtained as follows. Let $\mathfrak{A}$ be the CAR algebra over $\mathfrak{K} = L^2(0,\infty)$. Specifically, let $\mathfrak{K} = L^2(0,\infty)$ with inner product

$$(f,g) = \int_{-\infty}^{\infty} \overline{f(x)}g(x)\, dx.$$

Let $\mathfrak{H}_n = \mathfrak{K} \wedge \mathfrak{K} \wedge \cdots \wedge \mathfrak{K}$ (n-times) be the antisymmetric tensor product of $\mathfrak{K}$ with itself n times. We define $\mathfrak{H}_o = \mathbb{C}$ a one dimensional Hilbert space. Let $\mathfrak{H} = \oplus_{n=0}^{\infty} \mathfrak{H}_n$. This space is called Fock space and the space $\mathfrak{H}_n$ are called the n-particle spaces. For each $f \in \mathfrak{K}$ we define the bounded linear operator $a(f)$ on $\mathfrak{H}$ where $a(f)$ maps $\mathfrak{H}_n$ into $\mathfrak{H}_{n-1}$ by the relation

$$(a(f)F_n)(x_1,\ldots,x_{n-1}) = \sqrt{n} \int_0^{\infty} F_n(x_1,\ldots,x_{n-1},x_n)\, f(x_n)\, dx_n$$

One finds the adjoint operator $a(f)^*$ maps $\mathfrak{H}_n$ into $\mathfrak{H}_{n+1}$ as follows,

$$(a(f)^*F_n)(x_1,\ldots,x_n,x_{n+1}) =$$

$$(n+1)^{-\frac{1}{2}} \sum_{k=1}^{n+1} (-1)^{n+k+1}\; F(x_1,\ldots,x_{k-1},x_{k+1},\ldots,x_{n+1})\, \overline{f(x_k)}$$

The operators $a(f)$ are bounded and satisfy the CAR relations

i. $a(\alpha f + g) = \alpha a(f) + a(g)$

ii. $a(f)a(g) + a(g)a(f) = 0$

iii. $a(f)^*a(g) + a(g)a(f)^* = (f,g)\, I$

for all $f,g \in \mathfrak{K}$ and complex numbers α.

Let $\mathfrak{A}$ be the C^*-algebra generated by the a(f) with $f \in \mathcal{K}$. The algebra $\mathfrak{A}$ is called the CAR algebra over $\mathcal{K}$ and the particular realization of $\mathfrak{A}$ as operators on the Hilbert space $\mathfrak{H}$ is known as th Fock representation of $\mathfrak{A}$. A particular feature of the Fock representation which characterizes this representation up to unitar equivalence is the existence of a cyclic vector $F_o \in \mathfrak{H}_o$ (called the vacuum vector) which has the property that $a(f)F_o = 0$ for all $f \in \mathcal{K}$.

Let S(t) be the translation operator on $\mathcal{K}$ given by

$$
\begin{aligned}
(S_t f)(x) &= f(x - t) \quad \text{for } x \geq t \\
&= 0 \quad \text{for } x < t.
\end{aligned}
$$

We define $\alpha_t(a(f)) = a(S_t f)$ and $\alpha_t(p(a(f_1), \cdots, a(f_n)^*) = p(a(S_t f_1), \cdots, a(S_t f_n)^*)$ where p is a polynomial in the a(f) and the $a(f)^*$. Since the mapping $a(f) \longrightarrow a(S_t f)$ preserves the CAR relations we have that α_t extends to a continuous $*$-endomorphism of $\mathfrak{A}$ into itself. For each $t > 0$ we can extend α_t to all of $\mathfrak{B}(\mathfrak{H})$ by weak continuity. One checks that α_t is an E_o-semigroup of $\mathfrak{B}(\mathfrak{H})$.

Intuitively one feels that the index of this E_o-semigroup shou be one. The semigroup S_t is generated by an operator with deficien indices (0,1). So our problem is how to associate the number one with this E_o-semigroup.

We begin by noting that for this example there exists a strongly continuous one parameter semigroup of intertwining isometries (i.e., there is a strongly continuous semigroup of isometries U(t) so that $U(t)A = \alpha_t(A)U(t)$ for all $A \in \mathfrak{B}(\mathfrak{H})$ and $t \geq 0$). The intertwining isometries U(t) act on the n-particle spaces by the relations

$$U(t)F_n)(x_1,\ldots,x_n) = F_n(x_1 - t,\ldots,x_n - t) \text{ for } x_i \geq t \quad i = 1,\ldots,n$$
$$= 0 \quad \text{otherwise.}$$

-d be the generator of U(t). More specifically we define

$$df = \lim_{t\to 0^+} (f - U(t)f)/t$$

re the domain $\mathfrak{D}(d)$ of d is the set of all $f \in \mathfrak{H}$ so that the it exists in the sense of norm convergence. Let d^* be the mitian adjoint of d. Note that since the U(t) are isometric s skew-hermitian so $-d^*$ is an extension of d (i.e. $-d^* \supset d$).

Then it follow from the theory of hermitian operators (see . Dunford and Schwartz [3] Chapter XII section 4) that each $\mathfrak{D}(d^*)$ can be uniquely expressed in the form $f = f_o + f_+ + f_-$ where $_\pm = \pm f_\pm$ and $f_o \in \mathfrak{D}(d)$. There are no solutions to the equation $_- = -f_-$ since this would imply for $g \in \mathfrak{D}(d)$ that

$$\frac{d}{dt}(f_-,U(t)g) = -(f_-,dU(t)g) = -(d^*f_-,U(t)g) = (f_-,U(t)g).$$

ce, $(f_-,U(t)g) = e^t(f_-,g)$ for $g \in \mathfrak{D}(d)$ and this contradicts the t that the U(t) are isometric. Hence, each $f \in \mathfrak{D}(d^*)$ can be quely expressed in the form $f = f_o + f_+$ with $f_o \in \mathfrak{D}(d)$ and $f_+ \in \mathfrak{D}(d\ )$ h $d^*f_+ = f_+$. Note that the space of such vectors f_+ is precisely space $\mathfrak{M}$ of vectors $f \in \mathfrak{H}$ so that $U(t)^*f_+ = e^{-t}f_+$ for $t > 0$.

We define a bilinear form < , > on $\mathfrak{D}(d^*)$ as follows

$$\langle f,g\rangle = \tfrac{1}{2}(f,d^*g) + \tfrac{1}{2}(d^*f,g).$$

traight forward computation show that if $f = f_o + f_+$ and $g_o + g_+$ with $f_o,g_o \in \mathfrak{D}(d)$ and $f_+,g_+ \in \mathfrak{M}$ then $\langle f,g\rangle = (f_+,g_+)$.

Hence, the bilinear form $< , >$ is positive on $\mathfrak{D}(d^*)$.

Let δ be the generator of α_t. Specifically we define

$$\delta(A) = \lim_{t \to 0^+} (\alpha_t(A) - A)/t$$

where the domain $\mathfrak{D}(\delta)$ is the set of all $A \in \mathfrak{B}(\mathfrak{H})$ so that the above limit exists in the sense of σ-weak convergence. One easily checks that $\mathfrak{D}(\delta)$ is a σ-weakly dense $*$-subalgebra of $\mathfrak{B}(\mathfrak{H})$ and δ is a closed $*$-derivation of $\mathfrak{D}(\delta)$ into $\mathfrak{B}(\mathfrak{H})$.

First we will show that if $A \in \mathfrak{D}(\delta)$ then $A\mathfrak{D}(d) \subset \mathfrak{D}(d)$ and $dAf = -\delta(A)f + Adf$. To this end suppose $A \in \mathfrak{D}(\delta)$ and $f \in \mathfrak{D}(d)$. Then we have

$$\begin{aligned} t^{-1}(I - U(t))Af &= t^{-1}(I - \alpha_t(A)U(t))f \\ &= -t^{-1}(\alpha_t(A) - A)U(t)f + t^{-1}A(I - U(t))f. \\ &\to -\delta(A)f + Adf \quad \text{as } t \to 0^+. \end{aligned}$$

Hence, $Af \in D(d)$ and $dAf = -\delta(A)f + Adf$. Now suppose $f \in \mathfrak{D}(d^*)$, $A \in \mathfrak{D}(\delta)$ and $g \in \mathfrak{D}(\delta)$ then we have

$$(Af,dg) = (f,A^*dg) = (f,dA^*g) + (f,\delta(A^*)g) = ((Ad^*f + \delta(A)f),g).$$

Hence, we have for $A \in \mathfrak{D}(\delta)$ that $A\mathfrak{D}(d^*) \subset \mathfrak{D}(d^*)$ and for $f \in \mathfrak{D}(d^*)$

$$d^*Af = Ad^*f + \delta(A)f. \qquad (*)$$

Using $(*)$ we will show that the mapping $A \to A$ gives us a $*$-representation of $\mathfrak{D}(\delta)$ on $\mathfrak{D}(d^*)$ with respect to the bilinear form $<\cdot,\cdot>$. To this end suppose $f,g \in \mathfrak{D}(d^*)$ and $A \in \mathfrak{D}(\delta)$. Then we have

$$<f,Ag> = \tfrac{1}{2}(d^*f,Ag) + \tfrac{1}{2}(f,d^*Ag) = \tfrac{1}{2}(A^*d^*f,g) + \tfrac{1}{2}(f,Ad^*g) + \tfrac{1}{2}(f,\delta(A)g)$$

$$= \tfrac{1}{2}(d^*A^*f,g) + \tfrac{1}{2}(A^*f,d^*g) = <A^*f,g>.$$

If $f \in \mathfrak{D}(d^*)$ and $<f,f> = 0$ then $f \in \mathfrak{D}(d)$. Hence, the mapping $A \longrightarrow A$ gives a $*$-representation of $\mathfrak{D}(\delta)$ on the quotient space $\mathfrak{D}(d^*)$ mod $\mathfrak{D}(d)$ with inner product $<\cdot,\cdot>$. We will call this representation of $\mathfrak{D}(\delta)$ on this quotient space π_α.

The question arises, what is this representation? To see this consider the vector $F_1 = \sqrt{2}\, e^{-x} \in \mathfrak{H}_1$ (the one particle space). A straight forward computation shows that $<a(f)F_1,a(f)F_1> = 0$ for all $f \in \mathfrak{D}_o = \{f \in L^2(0,\infty)$ with $df/dx \in L^2(0,\infty)$ and $f(0) = 0\}$. It follows that $<F_1,p(a(f_1),\cdots,a(f_n)^*)F_1> = (F_o,p(a(f_1),\cdots,a(f_n)^*)F_o)$ for all polynomials p in the $a(f)$ and $a(f)^*$ with $f \in \mathfrak{D}_o$. Hence, the state associated with F_1 for the new representation π_α is the Fock state. A little work shows that (see [5] for details) shows that F_1 is a cyclic vector for π_α. Hence, π_α is unitarily equivalent to the Fock representation.

Now see why the index of this representation is one. We define the index as the multiplicity of π_α. Since π_α is irreducible π_α has multiplicity one.

Now suppose we repeat the above construction only this time we replace $\mathfrak{K} = L^2(0,\infty)$ by $K = L^2(0,\infty) \oplus \cdots\cdot \oplus L^2(0,\infty)$ (n times). Then $\mathfrak{K}$ is the space of n-component functions on $[0,\infty)$. If we repeat the above construction and form the inner product $<\cdot,\cdot>$ and construct the representation π_α we now find there are n functions $F^{(i)} \in \mathfrak{H}_1$ (the one particle space) given by $F_j^{(i)}(x) = \delta_{ij}\ \sqrt{2}\, e^{-x} \in \mathfrak{H}_1$ and we have $<F^{(i)},p(a(f_1),\cdots,a(f_m)^*)F^{(j)}) = \delta_{ij}(F_o,p(a(f_1),\cdots,a(f_m)^*)F_o)$. We

see then that π_α is the direct sum of n copies of the Fock representation. Then we say this E_0-semigroup is of index n. For future reference we will call these examples the CAR flows of rank

Given any E_0-semigroup α_t of $\mathfrak{B}(\mathfrak{H})$ with an intertwining semigro of isometries U(t) so $U(t)A = \alpha_t(A)U(t)$ one can define the inner product $<\cdot,\cdot>$ and construct the representation π_α. We should note that π_α is insensitive to inner perturbations. If α_t is an E_0-semigroup of $\mathfrak{B}(\mathfrak{H})$ with generator δ we can define a new generator δ_1 given by $\delta_1(A) = \delta(A) + [h,A]$ where $h \in \mathfrak{B}(\mathfrak{H})$ is skew hermitian. Then δ_1 is a generator of an E_0-semigroup β_t of $\mathfrak{B}(\mathfrak{H})$. If $-d$ is the generator of U(t) then we define $d_1 = d - h$ then we find $-d_1$ is the generator of a semigroup W(t) which intertwines β_t so that $W(t)A = \beta_t(A)W(t)$ for all $A \in \mathfrak{B}(\mathfrak{H})$. The bilinear form $<\cdot,\cdot>_1$ one construct from d_1 is equal to the bilinear form one constructs from d. Hence the representation π_α is unchanged by inner perturbations. The representation π_α is then something which classifies α_t up to "oute conjugacy".

We define the index of α_t to be the multiplicity of π_α. One can show the following (see [5] for details). Suppose α_t and β_t ar E_0-semigroups of $\mathfrak{B}(\mathfrak{H})$ with intertwining semigroups $U_i(t)$ for $i = 1,$ Suppose α_t and β_t are of indices i_1 and i_2. Let $\gamma_t = \alpha_t \otimes \beta_t$ be the tensor product of α_t and β_t where $\gamma(A \otimes B) = \alpha_t(A) \otimes \beta_t(B)$. Then $U(t) = U_1(t) \otimes U_2(t)$ is an intertwining semigroup for γ_t and one ca show that the index i of γ_t satisfies $i \leq i_1 + i_2$. (I expect the equality sign holds.)

The index theory I have outlined is based on two assumption. One is the existence of an intertwining semigroup U(t) of isometrie

The second assumption is the implicit assumption that the intertwining semigroup U(t) is unique. In my own work I considered this second assumption to be an obvious fact. It was not until I had been working on these problems for two years that I realized that even for the CAR flow the intertwining semigroup U(t) is not unique. In fact, there are a whole family of intertwining semigroups. I believe (because I have checked most but not all of the detail) that if one uses one the other intertwining semigroups one obtains the same representation and the same index.

Also I have found an example of a E_0-semigroup of $\mathfrak{B}(\mathfrak{H})$ for which there is no intertwining semigroup (see [6] for details). The example is constructed by taking a certain generalized free state of the CAR algebra. I will not discuss the details of the construction because I do not feel the detail are particularly important. What is important is that an intertwining semigroup need not exist.

When I arrived in Australia to visit Derek Robinson in October of 1986 I knew that intertwining semigroups U(t) need not exist and if they did exist they were most probably not unique. It was clear that we needed an approach to index theory that was independent of an intertwining semigroup. Here is an approach that Derek Robinson and I found. The theorems that follow can be found in [7].

The basic idea is the following. Suppose A and B are hermitian operators of deficiency indices (0,n) and (0,m) and you want to determine if n = m. Consider the operator $C = A \oplus (-B)$. If C has a deficiency indices (m,n) and C has self adjoint extension if and only if n = m. We use the analogous construction for E_0-semigroups.

We begin by introducing the notion of paired E_0-semigroups.

Definition 2. Suppose α_t and β_t are E_o-semigroups of M_1 and M_2. We say α_t and β_t are paired denoted $(\alpha_t, \beta_t) \in \mathfrak{P}$ if there exists a continuous one parameter group Γ_t of $*$-automorphisms of $M_1 \otimes M_2$ so that $\Gamma_t(A \otimes I) = \alpha_t(A) \otimes I$ and $\Gamma_{-t}(I \otimes B) = I \otimes \beta_t(B)$ for all $A \in M$ $B \in M_2$ and $t \geq 0$.

From the definition it follows that.

Theorem 3. Suppose $\alpha_t, \beta_t, \gamma_t$ and σ_t are E_o-semigroups of von Neumann algebras.

If $(\alpha_t, \beta_t) \in \mathfrak{P}$ and $(\gamma_t, \sigma_t) \in \mathfrak{P}$ then $(\alpha_t \otimes \gamma_t, \beta_t \otimes \sigma_t) \in \mathfrak{P}$.

If $(\alpha_t, \beta_t) \in \mathfrak{P}$ and α_t and γ_t are inner conjugate then (γ_t, β_t)

If $(\alpha_t, \beta_t) \in \mathfrak{P}$ and σ_t is a continuous one parameter group of $*$-automorphisms then $(\alpha_t \otimes \sigma_t, \beta_t) \in \mathfrak{P}$.

One can show

Theorem 4. Suppose α_t is an E_o-semigroup of $M = \mathfrak{B}(\mathfrak{H})$. Then there exists an E_o-semigroup β_t of $N = \mathfrak{B}(\mathfrak{H})$ so that $(\alpha_t, \beta_t) \in \mathfrak{P}$.

All E_o-semigroups of $\mathfrak{B}(\mathfrak{H})$ that are known to us we have the property that they are paired with themselves. It would be interesting to know if every E_o-semigroup of $\mathfrak{B}(\mathfrak{H})$ is paired with itself.

Theorem 5. Suppose $\alpha_t, \beta_t, \gamma_t$ and σ_t are E_o-semigroups of M_1, M_2, M_3 and M_4 and each of the M_i are $*$-isomorphic with $\mathfrak{B}(\mathfrak{H})$. Suppose $(\alpha_t, \beta_t) \in \mathfrak{P}$, $(\beta_t, \gamma_t) \in \mathfrak{P}$ and $(\gamma_t, \sigma_t) \in \mathfrak{P}$. Then $(\alpha_t, \sigma_t) \in \mathfrak{P}$.

With this theorem we can define the notion of an index for E_o-semigroups of $\mathfrak{B}(\mathfrak{H})$.

efinition 6. We say that two E_o-semigroups α_t and β_t of $\mathfrak{B}(\mathfrak{H})$ ave the same index if there is an E_o-semigroup σ_t of $\mathfrak{B}(\mathfrak{H})$ so that $\alpha_t,\sigma_t) \in \mathcal{P}$ and $(\beta_t,\sigma_t) \in \mathcal{P}$ (i.e., α_t and β_t have the same index if hey can both be paired with the same E_o-semigroup).

It follow immediately from theorem 5 that if α_t and β_t are $_o$-semigroups of the same index and β_t and γ_t are of the same index hen α_t and γ_t are of the same index. The notion of index then ivides the set of E_o-semigroups of $\mathfrak{B}(\mathfrak{H})$ into equivalence classes.

The obvious question is can this new index distinguish the CAR lows of rank n from one another. The answer is it can.

heorem 7. The CAR flow of rank n and the CAR flow of rank m have he same index if and only if n = m.

The proof of this theorem runs as follows. The CAR flows are onstructed using the CAR algebra. There is an analogous onstruction using the CCR algebra, the algebra of the canonical ommutation relations. In [7] we show that the E_o-semigroups onstructed from the CCR algebra are conjugate to the CAR flows. hen the above theorem is expressed in terms of the CCR algebra the he result can be obtain using the theory of Araki and Wood of ontinuous tensor products of Hilbert space [1].

heorem 8. Suppose α_t is an E_o-semigroup of $\mathfrak{B}(\mathfrak{H})$ and there exists strongly continuous semigroup of isometries U(t) so that $(t)A = \alpha_t(A)U(t)$ for all $A \in \mathfrak{B}(\mathfrak{H})$ and $t \geq 0$. Suppose β_t is an $_o$-semigroup of $\mathfrak{B}(\mathfrak{H})$ and $(\alpha_t,\beta_t) \in \mathcal{P}$. Then there is a strongly ontinuous semigroup of isometries V(t) so that for $V(t)A = \beta_t(A)V(t)$ or all $A \in \mathfrak{B}(\mathfrak{H})$ and $t \geq 0$.

Since there is an E_o-semigroup α_t of $\mathfrak{B}(\mathfrak{H})$ for which there is no strongly continuous intertwining semigroup of isometries (i.e., there is no strongly continuous semigroup of isometries $U(t)$ so that $U(t)A = \alpha_t(A)U(t)$ for all $A \in \mathfrak{B}(\mathfrak{H})$ and $t \geq 0$). Since for CAR flows there exists a strongly continuous semigroup of intertwining isometries it follows that any flow which can be paired to a CAR flow has a strongly continuous semigroup of intertwining isometries by the previous theorem. And, therefore, any flow which is of the same index as a CAR flow must necessarily have a strongly continuous semigroup of intertwining isometries. Hence, α_t is not of the same index as any CAR flow.

We see then that the set of indices for E_o-semigroups of $\mathfrak{B}(\mathfrak{H})$ corresponds to a set which properly contains the non negative integers plus a point at infinity.

References

1. Araki, H. and Woods, E.J.: *Complete Boolean Algebras of Type I Factors*, Pubs. Res. Inst. Math. Sciences Kyoto University, Series A Vol II (1966) 157-242.

2. Bratteli, O. and Robinson, D.: Operator Algebras and Quantum Statistical Mechanics I. Springer - Verlag, 1979.

3. Dunford, N. and Schwartz, J.: Linear Operators part II. Interscience, 1963.

4. Haag, R. and D. Kastler, *An algebraic approach to quantum field theory*, J. Math Phys. **7** (1964),848-861

5. Powers, R. T.; *An index theory for semigroups of *-endomorphisms of $\mathfrak{B}(\mathfrak{H})$ and type II_1 factors*, to appear in Canadian Jour. of Math.

6. Powers, R. T.: *A non spatial continuous semigroup of *-endomorphisms of* $\mathfrak{B}(\mathfrak{H})$, to appear in Pubs. Res. Inst. Math. Sciences Kyoto University.

7. Powers, R. T. and Robinson, D. W.; *An index for continuous semigroups of *-endomorphisms of $\mathfrak{B}(\mathfrak{H})$*, to appear in Jour. of Functional Analysis.

Coactions and Yang–Baxter Equations for Ergodic Actions and Subfactors

by Antony Wassermann
(University of Liverpool and University of California, Berkeley)

I. Introduction

The aim of this paper† is to bring into evidence the usefulness of considering not only actions of compact groups on operator algebras, but also the dual notion of coaction. A lot of what we shall say is contained in much greater detail in three series of papers, due to be published in the near future: four papers on ergodic actions, two on product type actions and two on equivariant K–theory. These all have their rather primitive origins in the three chapters of my thesis. The other main proponent of coactions of compact groups is Adrian Ocneanu, and we shall make frequent reference to his still unpublished work.

We now briefly summarise the contents of the rest of this paper. In Section II we recall the basic definitions of coactions of compact groups on von Neumann and C* algebras. We present two examples of C* algebras which arise perhaps unexpectedly as crossed products by coactions, and show how this observation can be used to explore their structure. The basic idea here is an old one: to use symmetry properties to simplify and elucidate computations. In Section III we exhibit two general principles in equivariant KK–theory, namely Frobenius Reciprocity and Dirac Induction. When combined with the equivariant Thom isomorphism, these lead to a generalisation of a spectral theorem of Hodgkin (for the K–theory of spaces) which in principle provides a homological machine whereby ordinary KK-theory (of a pair of algebras) can be deduced from equivariant KK–theory. In particular if G is a compact Lie group which is both connected and simply connected, this leads one to suspect that if A is a G–algebra for which $K_*^G(A)$ is just $\mathbb{Z}$ (with the augmentation action of $R(G)$), then A is KK–equivalent to $C(G)$. Examples of such actions arise naturally in the theory of ergodic actions.

In Section IV we outline the general theory of ergodic actions and show how they can be understood better by exploiting a link between equivariant K–theory and spectral multiplicities. This leads to the concept of the 'multiplicity map' and its associated diagrams. We explain how these ideas can be used to classify the ergodic actions of $SU(2)$. To achieve this we need two additional tools: the first exploits the theory of coactions, while the second hinges on the presence of gaps in the spectra of certain homogeneous spaces of $SU(2)$. We reach the perhaps disappointing conclusion that $SU(2)$ has no exotic ergodic actions

† This is an expanded version of a talk given at the U.S.–U.K. joint conference on Operator Algebras at Warwick in July, 1987.

on operator algebras, i.e. all its ergodic actions are necessarily on Type I von Neumann algebras. On the other hand, looking on the bright side, we recapture the classification of the closed subgroups of $SU(2)$ up to conjugacy by a new if somewhat long–winded method. In Section V we explain how it is possible to develop a theory which parallels in every respect the easy classification of ergodic actions of compact Abelian groups, but this time for non–commutative groups. The natural restriction here is that the crossed product by the ergodic action should be a factor. One finds that such actions are classified by cocycles and bicharacters of the group dual, plus analogues of the usual non–degeneracy criteria for the action to be on a factor. The non–commutativity, however, introduces a quite novel feature, the symmetric quantum Yang–Baxter equation. This equation (with parameters) has been much studied in Russia in connection with completely integrable Hamiltonian systems and exactly solved models in quantum field theory. It also turns out to be the key to understanding the second cohomology of the dual of a simple compact group. In the remainder of this section we illustrate how these cocycles can be classified for certain classical groups of low rank.

In Section VI we study actions of compact groups on von Neumann algebras for which the crossed product is a factor. We consider in particular actions where there is an invariant subalgebra satisfying the same hypothesis. We then examine the position of the fixed point algebra of the subalgebra in the fixed point algebra of the whole algebra using the framework provided by the theory of subfactors. We find that there is a particularly simple 'invariance principle' operating here. Finally in Section VII, we bring together the ideas of the two preceding sections. Every solution of the symmetric quantum Yang–Baxter equation gives rise to a factor representation of the infinite symmetric group in the hyperfinite II_1 factor, and hence we naturally obtain a subfactor. We conjecture that the position of this subfactor is more or less equivalent to the information provided by the bicharacter and go some way to lending credibility to this idea, by showing the analogy with the theory developed in Section VI.

The style of this paper has been kept deliberately terse, primarily because of limitations of space and time, but also because I preferred to emphasise broad principles rather than concentrating on fussy details. The same restrictions also apply to the list of references provided, which has deliberately been kept short and consequently no doubt has many glaring omissions. More complete references can be found in the bibliographies of the papers referred to at the beginning of this introduction. The debt of gratitude I owe to other mathematicians would be hard to repay, in particular to Bill Arveson, Dick Kadison, Claude Schochet and Masamichi Takesaki; I would also like to thank Vaughan Jones, Adrian Ocneanu, Picu Voiculescu and Hans Wenzl for their help and encouragement.

II. Coactions of Compact Groups

Let G be a compact group with C* algebra $C^*(G) = \oplus_{\pi\in\hat{G}} End(V_\pi)$ and right von Neumann algebra $\mathcal{R}(G) = \overline{\bigoplus} End(V_\pi)$. If A is a C* or W* algebra, a *coaction* of G (or action of $\hat{G}$) is described by a *-isomorphism $\delta: A \to A \otimes \mathcal{R}(G)$ such that $(\iota \otimes \delta_G)\delta = (\delta \otimes \iota)\delta$, where $\delta_G: \mathcal{R}(G) \to \mathcal{R}(G) \otimes \mathcal{R}(G)$ is the canonical coaction $\rho(g) \mapsto \rho(g) \otimes \rho(g)$. If G is Abelian, this clearly amounts to giving an action of $\hat{G}$ on A. Otherwise if $\pi_1, \ldots, \pi_n$ are a *faithful* family of irreducible representations of G, then δ is uniquely specified by $\delta_{\pi_1}, \ldots, \delta_{\pi_n}$, the parts of δ going into the π_k components of $A \otimes \mathcal{R}(G)$. We do not wish to enter into the duality theory for coactions and actions (which is anyway rather easy for compact groups). Suffice it to say that for every action on A there is a canonical coaction on the crossed product $A \rtimes G$, which is given by $(A \otimes \mathcal{K}(L^2(G)))^G$ or $(A \otimes \mathcal{B}(L^2(G)))^G$ according to whether A is a C* or W* algebra. It is induced by the natural isomorphism $L^2(G) \otimes V \cong \dim(V) \cdot L^2(G)$ of G-spaces. Moreover one defines the crossed product $A \rtimes_\delta \hat{G}$ by the coaction to be the subalgebra of $A \otimes \mathcal{B}(L^2(G))$ generated by $\delta(A)$ and $C(G)$ with the natural dual action of G given by $Ad(\lambda)$, where λ denotes the left regular representation of G on $L^2(G)$. Later we shall be concerned with coactions of G on von Neumann algebras, but first let us give two non-trivial examples arising as crossed products by coactions. The main criterion we shall use to recognize that an algebra has this form, that is supports an action of G which is dual, will be that of M. Landstad, namely the presence of an equivariant copy of $C(G)$ in the algebra. In both examples that we give there is a common underlying principle, namely that the structure of an algebra can be determined more easily by exploiting its symmetry properties to the full.

Example 1: Zeroth Order Pseudodifferential Operators on a Compact Lie Group G. Let $\mathcal{U}(G)$ be the ring of left invariant differential operators on G, the universal enveloping algebra of $Lie(G)$. If $X_1, \ldots, X_n$ is an orthonormal basis of $Lie(G)$ with respect to a G-invariant inner product, then we may define the Laplacian $\Delta = -(X_1^2 + \cdots + X_n^2)$. It is central in $\mathcal{U}(G)$ and therefore yields a positive scalar in every non-trivial irreducible representation of G. Regarding all operators in $\mathcal{U}(G)$ as sitting inside $\prod End(V_\pi)$, we may convert the first order differential operators $X_1, \ldots, X_n$ (vector fields) into zeroth order pseudodifferential operators $T_k = X_k/(I + \Delta)^{1/2}$. The C* algebra of zeroth order pseudodifferential operators $\Psi_0(G)$ on G is by definition the C* algebra generated on $L^2(G)$ by these bounded operators and $C(G)$. It is therefore by definition the crossed product by a coaction on $\Psi_0^G(G)$, the C*algebra of left invariant zeroth order pseudodifferential operators generated by just the T_k's. This enables one to establish the fundamental symbol sequence

$$0 \to \mathcal{K}(L^2(G)) \to \Psi_0(G) \to C(G \times S(Lie(G)^*)) \to 0$$

by considering the much simpler invariant sequence

$$0 \to C^*(G) \to \Psi_0^G(G) \to C(S(Lie(G)^*)) \to 0$$

and then forming the crossed product with $\hat{G}$. The invariant sequence is easy to analyse using asymptotic properties of the representation theory of G and is particularly easy when G is a torus (and the invariant algebra is Abelian). For general G, the asymptotics have close links with results of V. Guillemin, W. Lichtenstein and H. Widom.

Example 2: Toeplitz Operators on the Unitary Groups. The unitaries U_n form the Šilov boundary of the open unit ball $D_n = \{x : x^*x < 1\}$ in $M_n(\mathbb{C})$. The topological boundary ∂D_n of D_n breaks up naturally into n distinct parts corresponding to the number of eigenvalues of x^*x equal to 1. Each part is an orbit $\mathcal{O}_k$ for the natural action of $U(n,n)$ on $\overline{D}_n$ and is a union of components isomorphic to D_k with $1 \leq k \leq n$, whose Šilov boundaries are naturally contained U_n. Thus $\mathcal{O}_k = \cup_{u,v \in U_n} uD_kv^*$ with the Šilov boundary of uD_kv^* given by uU_kv^*. Let $H^2(U_n) \subset L^2(U_n)$ be the Hardy space given by the L^2 boundary values of holomorphic functions in this ball. Thus $H^2(U_n)$ is just the closure of the space of polynomials $p(z_{11}, \ldots, z_{nn})$ restricted to U_n. Let $P: L^2 \to H^2$ be the Hardy space projection and define $\mathcal{T}(U_n)$ to be the C* algebra generated by the Toeplitz operators $T(\phi) = Pm(\phi)P$ where $\phi \in C(U_n)$ and $m(\phi)$ is the corresponding multiplication operator on $L^2(U_n)$. The analogue of this algebra was considered in the unbounded (upper half-plane) realisation of the unit ball by Paul Muhly and Jean Renault. Their approach was to analyse the algebra using 'groupoid' techniques, which essentially boiled down to using crossed products by $\mathbb{R}^n$. The Hardy space projection in their case was given by convolution by a certain distribution on $\mathbb{R}^n$ which in Fourier corresponded to multiplication by the characteristic function of a cone. The structure of the algebra was then obtained by a microlocal analysis of this distribution, an analysis which simply amounted to compactifying the cone according to its asymptotics. Our approach to unravelling the structure of $\mathcal{T}(U_n)$ exactly mirrors that of Muhly and Renault, except that the euclidean group $\mathbb{R}^n$ is replaced by the compact group U_n (or more precisely $U_n \times U_n$). The structure can be described quite simply in terms of the boundary components of the unit ball and the analysis proceeds by using coactions and a sort of microlocal analysis on U_n. In fact from the Cauchy formula expressing the values of a holomorphic function in terms of its boundary values on U_n

$$f(z) = \int_{U_n} \frac{f(u)du}{\det(1 - u^*z)^n},$$

one can write $P(f) = \theta_n * f$ where θ_n is the central distribution $\det(1-u)^{-n} = \lim_{\varepsilon \uparrow 1} \det(1-\varepsilon u)^{-n} du$ on U_n. Let $\mathcal{A}_n$ be the C* algebra generated by the convolution operators $\phi \cdot \theta_n$ ($\phi \in C^\infty(U_n)$). This could be called the 'microlocal algebra' of θ_n, since its asymptotics (i.e. its irreducible representations) are very much related to the microlocal strucure of θ_n. In fact by simple representation theoretic arguments one can show that the irreducible representations of $\mathcal{A}_n$ come from the natural homomorphisms

$$\phi \cdot \theta_n \to \phi^g \cdot \theta_k \quad (\phi^g = \phi \text{ conjugated by } g)$$

associated with Hardy space distributions for smaller boundary components. The existence of these homomorphisms lies at the heart of our analysis and follows from the fact that if ψ_f is the highest weight matrix coefficient of U_n with signature $f : f_1 \geq f_2 \geq \cdots \geq f_n$, then

$$\lim_{N_1, \ldots, N_k \to \infty} \overline{\psi}_{N_1, \ldots, N_{n-k}, 0, \ldots, 0, \ldots, 0} \cdot \theta_n = \mathrm{ind}_{U_k \uparrow U_n} \theta_k.$$

This limit formula is established by using some elementary representation theory of U_n. The structure of $\mathcal{T}(U_n)$ (which is a $U_n \times U_n$–algebra) can then be deduced immediately once one establishes that it is isomorphic to $P(\mathcal{A}_n \rtimes \hat{U}_n)P$. This identification, however, is a direct consequence of the classical fact that the natural representation of $\mathcal{T}(U_n)$ on $H^2(U_n)$ is irreducible. Having performed this analysis, one then finds that the irreducible representations of $\mathcal{T}(U_n)$ are completely described by Toeplitz representations on the various boundary components of the unit ball. Thus one fixes $1 \leq k \leq n$ and $u, v \in U_n$ and defines an irreducible representation $\pi_{k,u,v}$ of $\mathcal{T}(U_n)$ by $T(\phi) \mapsto T(\phi|_{uU_kv^*})$, where as above uU_kv^* is regarded as the Šilov boundary of the unit ball uD_kv^*. As a consequence of this analysis one can apply an argument of Coburn to show that if $n \geq 2$ and f is a holomorphic function on D_n extending to a continuous function on $\overline{D}_n$, then the set of values taken by f on the topological boundary ∂D is equal to the closure of the set of values taken by f in D_n. This of course is a considerable strengthening of the usual maximal modulus principle.

Before leaving this example, we should point out that another line of attack on $\mathcal{T}(U_n)$ using Jordan triple systems was pursued independently by Harald Upmeier, who succeeded in obtaining a uniform classification for the Toeplitz algebras corresponding to all of E. Cartan's bounded symmetric domains. It is not clear whether the remaining domains, namely those not of type I_{nn}, can be dealt with by an adaptation of our 'non–commutative groupoid' approach, although this does not seem totally implausible.

III. Equivariant Kasparov Theory
Frobenius Reciprocity, Dirac Induction and Hodgkin's Spectral Sequence

The C* algebra $\mathcal{T}(U_n)$ we considered in the last example has a sequence of $U_n \times U_n$–invariant ideals

$$\mathcal{T}(U_n) \supset J_1 \supset J_2 \supset \cdots \supset J_n \supset \{0\}$$

with successive quotients $J_k/J_{k+1} \cong ind_{N(U_k)\uparrow U_n \times U_n} \mathcal{K}(L^2(U_k))$ as $U_n \times U_n$–algebras. It is therefore of great interest to study the extensions

$$0 \to J_{k+1}/J_{k+2} \to J_k/J_{k+2} \to J_k/J_{k+1} \to 0$$

as elements of the *equivariant Kasparov group* $KK^1_{U_n \times U_n}(J_k/J_{k+1}, J_{k+1}/J_{k+2})$ or even as elements of the ordinary Kasparov group $KK^1(J_k/J_{k+1}, J_{k+1}/J_{k+2})$. Upmeier has used the index theorems of Boutet de Monvel to identify these elements explicitly, but another method is available if one exploits the symmetries a little. In fact there are two basic principles in equivariant KK–theory:

1. *Frobenius Reciprocity:* $KK_G(A, ind_{H\uparrow G}B) \cong KK_H(res_{G\downarrow H}A, B)$ whenever H is a closed subgroup of G with A a G–algebra and B an H–algebra.
2. *Dirac Induction:* $KK_T(A, B) \cong R(T) \otimes_{R(G)} KK_G(A, B)$ whenever T is a maximal torus in the compact Lie group G and A and B are G–algebras. We must assume in addition that G is connected and that $\Pi_1(G)$ is torsion–free.

The first of these results is an example of the metamathematical principle that anything functorially true for $Hom_G(A,B)$ should extend automatically to $KK_G(A,B)$. This is because elements of the latter group are morally just differences of homomorphisms. A precise form of this vague statement has been supplied by J. Cuntz's alternative description of KK–theory via quasi–homomorphisms and (more recently) his algebra qA. Our proof uses an imprimitivity theorem for induced Hilbert C* modules, but it would be interesting to supply a more direct proof based on an equivariant version of qA. The second result on Dirac induction is much harder to prove. The main ingredients of the proof are as follows:

(i) Steinberg's Basis Theorem and the formalism of the discrimant
(ii) Bott's Index Theorem for homogeneous elliptic operators on G/T
(iii) Borel's T–equivariant cell decomposition of G/T
(iv) Weyl's Character Formula

Let us give a brief sketch of the proof. By (i) one knows that $R(T)$ is a free $R(G)$–module of rank $|W|$ where $W = N(T)/T$ is the Weyl group. Moreover Steinberg actually constructs an explicit basis given by characters in $\hat{T}$. Thus the tensor product $R(T)\otimes_{R(G)} KK_G(A,B)$ may be identified with the direct sum of $|W|$ copies of $KK_G(A,B)$ as an $R(G)$–module. Our task is therefore to construct maps $\alpha: KK_T(A,B) \to KK_G(A,B)^{|W|}$ and $\beta: KK_G(A,B)^{|W|} \to KK_T(A,B)$ such that $\alpha\beta$ and $\beta\alpha$ are the identity maps. To do so, we use Frobenius Reciprocity to identify $KK_T(A,B)$ with $KK_G(A,B\otimes C(G/T))$. We then produce elements $x_1,\ldots,x_{|W|}$ in $KK_G(\mathbb{C},C(G/T))$ and elements $y_1,\ldots,y_{|W|}$ in $KK_G(C(G/T),\mathbb{C})$ which form 'dual bases' in the sense that (1) $x_i\otimes_{\mathbb{C}} y_j = \delta_{ij}$ and (2) $\sum_{i=1}^{|W|} y_i \otimes_{C(G/T)} x_i = \tau_{C(G/T)}(c_1)$. Here we take Kasparov products and the element on the right hand side of the second equation is just the identity element of the ring $KK_G(C(G/T),C(G/T))$. The isomorphisms α and β can then be defined using Kasparov products as $\alpha(\xi) = (\xi\otimes_{C(G/T)}y_j)$ and $\beta(\eta_j) = \sum_i \eta_j\otimes_{\mathbb{C}}x_j$ where $\xi\in KK_G(A,B\otimes C(G/T))$ and $(\eta_j)\in KK_G(A,B)^{|W|}$.

Thus everything hinges on constructing the dual bases and in fact the main work lies in producing a solution of (1). By taking the x_i's as homogeneous line bundles E_i over G/T and the y_j's as Dirac operators D_{F_j} on G/T with coefficents in appropriate homomogeneous line bundles F_j, the proof that the first equation can be satisfied reduces to the computation of a determinant. Indeed the E_i's and F_j's essentially correspond to characters of T. Moreover the Kasparov product $x_i\otimes_{C(G/T)}y_j$ is given by the $R(G)$–valued index of $D_{E_i\otimes F_j}$. On the other hand, by (ii) and (iv), the index of D_E is equal to $A(E\otimes E_0)/A(E_0)$ where E_0 is a certain fixed character of T and $A(E)$ denotes the alternating sum $\sum_{\sigma\in W}(-1)^\sigma\sigma(E)$ over the Weyl group. Thus to solve (1) it suffices to find characters $E_1,\ldots,E_{|W|}$ such that the determinant $\Delta = \det(A(E_i\otimes E_j))$ is equal to $\pm A(E_0)^{|W|}$. Let $\sigma_1,\ldots,\sigma_{|W|}$ be the elements of W and consider the matrices $M = (\sigma_i(E_j))$ and $N = ((-1)^{\sigma_i}\sigma_i(E_j))$. We then have $\det(M)^2 = \pm\det(N^t)\det(M) = \pm\det(N^tM) = \pm\Delta$ so that up to a sign Δ is given by the square of $\det(M)$. On the other hand the Steinberg basis of $R(T)$ satifies $\det(M) = A(E_0)^{|W|/2}$, so we are through.

Let us now indicate how (1) and (iii) together imply (2). We have to show that the idempotent $e = \sum_i y_i \otimes_{\mathbb{C}} x_i$ is the identity in the ring $R = KK_G(C(G/T), C(G/T))$. It is easily verified that e is a two sided unit for elements of the form $y_i \otimes_{\mathbb{C}} x_j$, which themselves form a linearly independent set over $R(G)$. Let R_0 be the $R(G)$–submodule of R that they generate. We claim that (over $R(G)$) R is torsion free and R/R_0 is a torsion module. To see this we use Frobenius Reciprocity to identify R with $KK_T(C(G/T), \mathbb{C})$ (as an $R(G)$–module). The Borel cell decomposition for G/T and the T–equivariant Thom isomorphism then imply that $R = KK_T(C(G/T), \mathbb{C})$ is torsion–free as an $R(T)$–module and that if $k(T)$ is the field of fractions of $R(T)$, then the dimension of $k(T) \otimes_{R(T)} R$ over $k(T)$ is exactly $|W|$. (This idea goes back to J. McLeod.) Returning to the $R(G)$–module structure of R, we deduce that R is torsion–free over $R(G)$. Moreover some simple Galois theory shows that $k(G) \otimes_{R(G)} R$ has dimension $|W|^2$ over $k(G)$, where $k(G)$ is the field of fractions of $R(G)$. Since R_0 is free of rank $|W|^2$, it follows that R/R_0 is a torsion module. Let us now show that e is the identity element 1 of R. Since R/R_0 is a torsion module, we can certainly find a non–zero element $a \in R(G)$ such that $a1 \in R_0$. Hence $e(a1) = a1$ so that $a(e-1) = 0$. Since R has no torsion, we infer that $e = 1$ as required. (From this one can deduce that $R = R_0$ using the easily verified relation $R_0 R R_0 \subseteq R$ and hence that the elements $y_i \otimes_{\mathbb{C}} x_j$ form an $R(G)$–basis of $KK_G(C(G/T), C(G/T))$. A similar argument shows that the y_i's form a basis of $KK_G(C(G/T), \mathbb{C})$. In this way we obtain an alternative combinatorial approach to Poincaré duality for the homogeneous space G/T.)

Note that as a result of Frobenius Reciprocity one can identify the group $KK_{U_n \times U_n}(ind(\mathcal{K}(H^2(U_k))), ind(\mathcal{K}(H^2(U_{k+1}))))$ with the equivariant K-homology group $KK_{N(U_{k+1})}(ind(\mathcal{K}(H^2(U_k))), \mathbb{C})$ which is much easier to understand. We shall now indicate a much more useful application of the two principles. We start with the extension of $\mathbb{Z}$–algebras

$$0 \to C_0(\mathbb{R}\backslash\mathbb{Z}) \to C_0(\mathbb{R}) \to C_0(\mathbb{Z}) \to 0$$

where $\mathbb{Z}$ acts on $\mathbb{R}$ by translation. Passing to the n–fold tensor product and taking the crossed product by $\mathbb{Z}^n$, one obtains a natural filtration on the algebra $C_0(\mathbb{R}^n) \rtimes \mathbb{Z}^n$ which is compatible with the dual action of $\mathbb{T}^n$. This filtration appears implicitly in Kasparov's conspectus as a special case of a far more general construction. If $T = \mathbb{T}^n$ is a maximal torus in G then we obtain a corresponding filtration on the induced algebra $ind_{T\uparrow G} C_0(\mathbb{R}^n) \rtimes \hat{T}$. On the other hand, the $\mathbb{Z}$–equivariant Thom isomorphism theorem implies that this induced algebra is actually KK_G–equivalent to $C(G)$. Combining this observation with the spectral sequence corresponding to the filtration of the induced algebra immediately yields a spectral sequence converging to $KK_G(A, B \otimes C(G)) \cong KK(A, B)$. In this way we obtain a generalisation of Hodgkin's Spectral Sequence to equivariant KK–theory. Routine computations similar to those of Fack and Skandalis allow one to work out the E_2 term explicitly, the end result being a spectral sequence for arbitrary ($\mathbb{Z}_2$-graded) G–algebras A and B converging to $KK^*(A, B)$ with E_2 term $Tor^p_{R(G)}(KK^q_G(A, B), \mathbb{Z})$. Symbolically we have

$$Tor^p_{R(G)}(KK^q_G(A, B), \mathbb{Z}) \Rightarrow KK^{p+q}(A, B).$$

Thus in principle ordinary KK–theory can be deduced from equivariant KK–theory just by using homological algebra. Not only is the method of derivation we have outlined above significantly shorter than the original method of Hodgkin and Snaith, but it also yields a more general result. It is fair to say that this is largely due to the systematic use of Kasparov products. These already appeared implicitly in Atiyah's use of Dirac induction to obtain a splitting for the forgetful map $K_G(X) \to K_T(X)$, a result which was one of the key ingredients in Hodgkin's method. His method, however, relied strongly on techniques from algebraic topology for constructing spectral sequences. This had the disadvantage of restricting the class of spaces to which the result applied, a restriction which disappears if one works in the more flexible framework of Kasparov theory and uses 'elliptic topology' instead of 'algebraic topology'.

For the remainder of this paper, we shall not be concerned with general Kasparov theory which we have not really defined. Instead we shall only consider the groups $KK_G^*(\mathbb{C}, A) = K_*^G(A)$, that is the equivariant K–groups of A. Somewhat unexpectedly perhaps these will play an important rôle in the next seeemingly unrelated topic. By definition $K_0^G(A)$ is the Grothendieck of finitely generated projective A–modules $\mathcal{E}$ with compatible actions of G. It has a natural $R(G)$–module structure described by the maps $\mathcal{E} \mapsto V \otimes \mathcal{E}$ where V is a finite–dimensional representation of G. Now a basic result of P. Green and P. Julg states that $K_0^G(A)$ and $K_0(A \rtimes G)$ are canonically isomorphic. One can also see this by using Kasparov's equivalent Hilbert module formulation of $KK_G(\mathbb{C}, A)$ together with the natural identification of $A \rtimes G$ with $(A \otimes \mathcal{K}(L^2(G)))^{\alpha \otimes \mathrm{Ad}\lambda}$. More generally Baaj and Skandalis have obtained (as a special case of a result on actions of Hopf algebras on C* algebras) a canonical isomorphism $KK_G(A, B) \cong KK_{\hat{G}}(A \rtimes G, B \rtimes G)$ where $KK_{\hat{G}}$ is defined by a straightforward extension of Kasparov's definitions for Abelian G. Here one has to view the coaction as a module action of the Fourier algebra $A(G)$ of G, or even its G–finite part, i.e. the algebra of matrix coefficients of G. Their proof has two main ingredients: the Landstad–Takesaki duality theory for actions and coactions; and a simplified version of Higson's proof of Kasparov's technical lemma that is well adapted to coactions.

IV. Ergodic Actions of Compact Groups Equivariant K–theory and the Multiplicity Map

Let $\mathcal{M}$ be a von Neumann algebra and G a compact group acting on $\mathcal{M}$. This action is said to be *ergodic* if the fixed point algebra $\mathcal{M}^G$ is trivial, i.e. $\mathcal{M}^G = \mathbb{C}$. A complete classification of ergodic actions of compact Abelian groups was obtained first by Albeverio and Hoegh-Krohn and later by Olesen, Pedersen and Takesaki. The idea was very simple. We may assume the action of G faithful. If $u_\chi \neq 0$ is an eigenvector in the spectral subspace $\mathcal{M}_\chi$ of $\mathcal{M}$ corresponding to $\chi \in \hat{G}$, then $u_\chi u_\chi^*$ and $u_\chi^* u_\chi$ are both invariant under G so must be scalars, necessarily equal. Thus all non–zero spectral subspaces $\mathcal{M}_\chi$ are one–dimensional spanned by a unitary eigenvector u_χ. Since the set of

χ for which $\mathcal{M}_\chi \neq \{0\}$ is evidently a subgroup of $\hat{G}$, the faithfulness of the action of G forces it to be the whole of $\hat{G}$. It follows that the unique G–invariant state of $\mathcal{M}$ (given by averaging over G) is a trace. The assignment $\chi \mapsto u_\chi$ defines a projective representation of $\hat{G}$ corresponding to a unique class of cocycle in $H^2(\hat{G}, \mathbb{T})$. Due to the presence of the dual action of G, this is necessarily the regular representation of $\hat{G}$ on $L^2(\hat{G})$ (or $L^2(G)$ in Fourier) corresponding to the particular cocycle. This process can be reversed: in other words every faithful ergodic action of G is the natural action on the von Neumann algebra generated by some regular cocycle representation of $\hat{G}$. Thus one gets the natural generalisations of the non–commutative tori of Marc Rieffel. Physicists, however, would prefer G to be non–Abelian. Thus the basic open question is whether any non–Abelian compact Lie groups can admit non–trivial ergodic actions on von Neumann algebras, that is on the hyperfinite type II_1 factor. (The reduction to the factor case follows by using induced actions.) In the remainder of this section and the next section we will develop techniques which allow us to tackle this problem for groups of low rank, such as $SU(2)$ and $SU(3)$.

Let us now outline the basic elements in the general theory of ergodic actions and explain its position in the general programme for classifying actions of compact groups on von Neumann algebras. The first and most crucial result in the subject is the finiteness theorem of Høegh–Krohn, Landstad and Størmer which asserts that the unique G–invariant state ϕ on $\mathcal{M}$ (obtained by averaging over G) is in fact a trace and that every irreducible representation of G occurs with multiplicity less than or equal to its degree in the spectral decomposition $\mathcal{M} = \oplus_{\pi\in\hat{G}}\mathcal{M}_\pi$ of $\mathcal{M}$. This can be proved by completely elementary means by comparing the two inner products $\phi(xy^*)$ and $\phi(y^*x)$ on finite–dimensional G–submodules of $\mathcal{M}$ and using Hermann Weyl's dimension formula. One has to make constant use of *eigenmatrices* in the proof: for $\pi \in G$, we define an eigenmatrix for π to be an element $M \in \mathcal{M} \otimes End(V_\pi)$ such that $\alpha_g(M) = M\pi(g)$ for all $g \in G$. (One can also define the notion of an eigenvector for π as an element of $\mathcal{M} \otimes V^*$ satisfying the same condition as an eigenmatrix; thus the rows of an eigenmatrix M form eigenvectors for π and each eigenvector corresponds to a copy of π in $\mathcal{M}$.) The basic ideas in our proof of the 'finiteness' theorem nevertheless are exactly the original ones, although no direct appeal to Tomita theory is necessary. Once one has the result of the theorem, it follows that $(\mathcal{M} \otimes End(V))^G$ is finite–dimensional for any finite–dimensional representation V of G. The same is true if $\mathcal{M}$ is replaced by the C* algebra A generated by the spectral subspaces of $\mathcal{M}$. Thus $\mathcal{M} \rtimes G$ and $A \rtimes G$ have approximate identities consisting of finite rank projections. (We say that a projection is of finite rank if the corner it defines is finite–dimensional.) Thus $A \rtimes G \cong \oplus\mathcal{K}(\mathcal{H}_i)$ and $\mathcal{M} \rtimes G \cong \overline{\oplus}\mathcal{B}(\mathcal{H}_i)$ are both direct sums of algebras of compact operators or bounded operators. The main idea is then to arrange for a confrontation of the two distinct approaches suggested by the von Neumann theory and the C* theory. Before doing so, we can enlarge the problem slightly and notice that the study of ergodic actions is equivalent to the study of actions for which the crossed product $\mathcal{M} \rtimes G$ is type I with atomic centre and such that $\mathcal{M}^G$ contains no non–scalar central elements, i.e. $\mathcal{M}' \cap \mathcal{M}^G = \mathbb{C}$. (This is equivalent to the fact that G acts ergodically on the centre of $\mathcal{M}$.) This class of actions is stable under perturbations. Thus we are really classifying coactions of G on atomic type I algebras. Adrian Ocneanu has tackled the problem of classifying *prime* or *minimal* actions, namely those for which the crossed product $\mathcal{M} \rtimes G$ is a type II_∞ factor and for which the fixed point algebra has trivial relative commutant, $\mathcal{M} \cap (\mathcal{M}^G)' = \mathbb{C}$. This work remains unpublished, since Ocneanu has reason to believe that his machine for classifying subfactors of finite depth can be extended to yield a classification of prime actions. The end result is that all prime actions are conjugate.

Since it will be needed in the last two sections in this paper, we pause here to give the quintessential example of a prime action of an arbitrary compact Lie group G on the hyperfinite II_1 factor. Let π be a (projective) representation of G on a finite–dimensional vector space V such that the homomorphism $Ad\pi: G \to \mathcal{PU}(V) = Aut(End(V))$ is faithful. Let A be the UHF algebra $\otimes_{n=1}^{\infty} End(V)$ with its canonical trace τ and let $\mathcal{R}$ be the von Neumann algebra $\pi_\tau(A)''$. G acts on both these algebras by the infinite tensor action $\alpha_g = \otimes Ad(\pi(g))$. Now τ restricts to an extremal (or factor) trace on the fixed point algebra A^G. This follows essentially by an old observation of D. Voiculescu: the extremal traces of A^G correspond to the order–preserving homomorphisms of the ordered ring $K_0(A^G)$ into $\mathbb{R}$. Hence $\mathcal{R}^G$ must be a subfactor of $\mathcal{R}$. Now the group S_∞ of finite permutations of the positive integers has a natural representation π in A given by the canonical embeddings of the finite symmetric groups S_n in $\otimes_{i=1}^{n} End(V) = End(V^{\otimes n})$. Moreover $\pi(S_\infty)$ lies in $\mathcal{R}^G$ and actually generates the subfactor if $G = \mathcal{PU}(V)$ or $SU(V)$. It was Størmer who first observed that the relative commutant of $\pi(S_\infty)$ in $\mathcal{R}$ is trivial. Let us give a short proof of this which is useful in computing more general relative commutants. The idea is to show that $\lim_{n\to\infty} Av_{S_n}(x)$ exists in the weak topology of the Hilbert space $L^2(R,\tau)$ and is equal to $\tau(x)$. (Here Av_{S_n} denotes the average over the subgroup S_n of S_∞ consisting of permutations which fix the integers greater than n.) To prove this assertion, it clearly suffices to show for elements x and y of finite support in $\mathcal{R}$ (in an obvious sense) that $\tau(Av_{S_n}(x)y) \to \tau(x)\tau(y)$. This is a consequence of the easily verified observation that when n is large the supports of $\sigma(x)$ and y are disjoint for the majority of elements $\sigma \in S_n$, in which case $\tau(\sigma(x)y) = \tau(x)\tau(y)$. It remains to show that the crossed product is a factor. Since $\mathcal{R} \rtimes G$ is given by $(\mathcal{R} \otimes \mathcal{B}(L^2(G)))^{\alpha\otimes Ad\lambda}$, it will suffice to prove that $(\mathcal{R}\otimes End(W))^G$ is a factor for any finite–dimensional G–invariant subspace W of $L^2(G)$; for then the reductions by the projections onto increasing W's will be factor projections in the crossed product increasing strongly to the identity. On the other hand some elementary representation theory of $SU(V)$ shows that the representation W is contained in some tensor power $V^{\otimes nk}$ for k sufficiently large. Thus $(\mathcal{R} \otimes End(W))^G$ is a reduction of $(\mathcal{R} \otimes (End(V))^{\otimes nk})^G$, which is clearly isomorphic to $\mathcal{R}^G$. Hence $(\mathcal{R}\otimes End(W))^G$ is also a factor, as required. Using the same 'absorption' trick together with Connes' 2×2 matrix trick, one can also show that unitary eigenmatrices exist for all $\pi \in \hat{G}$. It turns out that this minimal action is also *dual*, that is there is a copy of $L^\infty(G)$ equivariantly embedded in R.

When G is Abelian this is a direct consequence of one of the main results of Ocneanu on the vanishing of 2–cohomology for centrally free actions of $\hat{G}$: it was Vaughan Jones who first pointed out that this immediately implies the essential uniqueness of prime actions of compact Abelian groups. When G is non–Abelian, Ocneanu has announced that a similar vanishing result holds for coactions of G (or even for centrally free actions of discrete Kac algebras). In the case of our product type action, however, it seems possible to give a direct asymptotic construction of $L^\infty(\mathcal{PU}(V))$ (and hence $L^\infty(G)$) using eigenmatrix techniques, although this procedure is of course not canonical. This construction is more or less equivalent to Ocneanu's construction of a model coaction of G. As we shall see in Section VII, the existence of such an embedding — or rather its 'quantum' analogues — is intimately tied up with the theory of ergodic actions. Ocneanu uses it to prove the uniqueness of prime actions of non–Abelian compact groups. There remain nevertheless many unanswered questions which may be considered fairly fundamental in the theory of automorphic actions of compact groups. For example for actions on a factor, must the crossed product have the same type as the fixed point algebra? It would be enough to provide an answer when the fixed point algebra is of the form $L^\infty(X)$ with X non–atomic. To my knowledge, nothing at all is known about this. Despite the existence currently of a vast industry devoted to the study of actions of compact groups on C* algebras, it seems that the more basic problems on von Neumann algebras have sadly been neglected.

Returning to ergodic actions, we start with the C* algebraic approach and don our K–theoretic cap. Thus we have $K_0^G(A) = K(A \rtimes G) = \oplus_{i=1}^n \mathbb{Z}$ where $1 \leq n \leq \infty$. Furthermore since we have a G–invariant trace $\tau \equiv \phi$ on A, we automatically get a dual trace $\tau_*: K_0^G(A) \to \mathbb{C}$ possessing the following compatibility with the $R(G)$–module structure of $K_0^G(A)$: $\tau_*(\pi \cdot e) = dim(\pi)\tau_*(e)$ for $\pi \in R(G)$ and $e \in K_0^G(A)$). Let us look at the von Neumann picture to interpret this result. Let e_i be the minimal projection in the ith component of $\mathcal{M} \rtimes G = (\mathcal{M} \otimes \mathcal{B})^G \cong \oplus_{i=1}^n \mathcal{B}_i$ (and hence the C* analogues). The e_i's are thus the generators of $K_0^G(A)$. τ_* corresponds to the trace on the crossed product obtained by restricting the obvious product trace on $\mathcal{M} \otimes \mathcal{B}$. Let $c_i = \tau_*(e_i)$, so that $0 < c_i < \infty$. The module action of $R(G)$ gives rise to a *-homomorphism $\mathbf{M}: R(G) \to M_n(\mathbb{Z})$ such that $\mathbf{M}(\pi)\mathbf{c} = dim(\pi)\mathbf{c}$. Our main result is then that $\mathbf{M}$ is a *multiplicity matrix*. Thus $M_{ij}(\pi)$ is the multiplicity of $\pi \in \hat{G}$ in $e_i(\mathcal{M} \otimes \mathcal{B})e_j$ and each of these multiplicities is bounded by the dimension of π. Thus by passing from the original ergodic action on $\mathcal{M} = e_1(\mathcal{M} \otimes \mathcal{B})e_1$ (say) to the associated (or conjugate) ergodic actions on $e_i(\mathcal{M} \otimes \mathcal{B})e_i$ together with the intertwiner spaces $e_i(\mathcal{M} \otimes \mathcal{B})e_j$, one finds the finds the remarkable property of *multiplicativity of multiplicities*. Data of the multiplicity matrices can be conveniently recorded on multiplicity diagrams. These are directed graphs with the $\mathbf{M}(\pi)$ as the incidences matrices (π need not be irreducible here). The condition $\mathcal{M}^G \cap \mathcal{M}' = \mathbb{C}$ is equivalent to the connectivity of the graph provided that π is faithful. It is a straightforward exercise in combinatorics to check that the only possible multiplicity diagrams for the two–dimensional identity representation of $SU(2)$ correspond exactly to the McKay diagrams for the closed subgroups of $SU(2)$ (that is — with the exception of the trivial subgroups — the extended

Dynkin diagrams of type $\tilde{A_n}$ $(1 \leq n \leq \infty)$, $\tilde{D_n}$ $(2 \leq n \leq \infty)$, $\tilde{E_6}$, $\tilde{E_7}$, and $\tilde{E_8}$ with their canonical weights). Thus from the point of view of spectral multiplicities, every ergodic action 'looks like' an action on $ind_{H\uparrow SU(2)} End(V)$ where V is a (projective) irreducible representation of a closed subgroup H of $SU(2)$. Broadly speaking there are two strategies for showing that this isomorphism of G–spaces can be made to respect the algebraic structure, thereby proving the main result on ergodic actions of $SU(2)$ — a 'coaction' strategy and a 'gap' stratgegy. In the first strategy one has to show that a certain action is dual, usually the action on a suitable reduction of $\mathcal{M} \otimes \mathcal{B}$. This reduction is chosen so that the fixed point algebra looks like a copy of the von Neumann algebra of the corresponding subgroup H, and in this case there are always unitary eigenmatrices. This works fine for the maximal torus and its subgroups. One shows that the action is dual by applying a general 'eigenmatrix criterion', which in this case amounts to exhibiting a unitary eigenmatrix for the two–dimensional representation having the form $\begin{pmatrix} a & b \\ -b^* & a^* \end{pmatrix}$. For a general connected compact Lie group, one knows that G is algebraic (in a particular faithful representation π) and the duality criterion is then that G should have a unitary eigenmatrix for π with commuting normal entries such that the algebraic equations defining G are satisfied by the entries of the matrix. The cases corresponding to the normaliser of the maximal torus and its binary dihedral subgroups are again handled by the coaction strategy, but this time it is necessary to pass to the fixed point algebra of the centre of $SU(2)$. Here one has an action of $SO(3)$ and one now looks for three–dimensional eigenmatrices with commuting self–adjoint entries and determinant one. The basic methods of producing such special eigenmatrices are illustrated in the first and third examples of the next section. The remaining subgroups H are treated by the gap strategy. One has an algebra $\mathcal{M}$ with the same multiplicities as $L^\infty(G/H)$ and in each case $L^\infty(G/H)$ has enough gaps in its 'G–spectrum' to force it to be Abelian. Thus for example one might have some irreducible representation π such that no subrepresentation of $\lambda^2\pi$ appeared in $L^\infty(G/H)$. The vectors in any copy of π in $\mathcal{M}$ would have to form a commuting set, as can be seen by considering the maps $\lambda^2 V \to \mathcal{M}$, $v \wedge w \mapsto vw - wv$, where V is some G–invariant subspace of $\mathcal{M}$. The gap strategy thus juxtaposes the internal product structure of the algebra with the external tensor product structure of representations. As an example let us consider an algebra $\mathcal{M}$ having the same spectral decomposition as $L^\infty(SU(2)/H)$ where H is the normaliser of a maximal torus. Thus the five–dimensional representation π_2 of $SU(2)$ appears with multiplicity one and, since it is real, it can be represented by a self–adjoint basis. Since no subrepresentation of π_2 occurs in $\mathcal{M}$, this basis generates an Abelian von Neumann algebra which is $SU(2)$–invariant. It is necessarily of the form $L^\infty(SU(2)/K)$ for some closed subgroup K and since π_2 is the smallest non–trivial representation to occur in it, one can easily conclude that K and H are conjugate subgroups of $SU(2)$. It follows therefore that $\mathcal{M}$ and $L^\infty(SU(2)/H)$ are equivariantly isomorphic as required.

V. Full Multiplicity Ergodic Actions of Compact Groups Bicharacters and Examples

As we have seen the classification of ergodic actions of compact Abelian groups reduced immediately to the study of *faithful* ergodic actions. It turns out that this condition is satisfied precisely when the crossed product is a factor. Analogously for non–Abelian G, there is an important class of ergodic actions to consider first before tackling the general case. These actions are characterised by the following equivalent conditions:

1. $dim(\mathcal{M}_\pi) = dim(\pi)$ for all $\pi \in \hat{G}$ (i.e. the action has full multiplicity)
2. each $\pi \in \hat{G}$ has a unitary eigenmatrix in $\mathcal{M}$
3. the W* crossed product is a factor (of type I)
4. the C* crossed product is isomorphic to an algebra of compact operators.

Thus in this case the multiplicity map is trivial because $K_0^G(A)$ is equal to $\mathbb{Z}$ with the augmentation action of $R(G)$. In view of Hodgkin's spectral sequence, it is not unreasonable to conjecture that A and $C(G)$ should be KK–equivalent or at least that their K–theories should coincide. Indeed one would expect the generators of the K–theory to appear in $K_1(A)$ as the unitary eigenmatrices of the 'fundamental' representations of G. This is a path we have not explored. Instead we have extended the cocycle and bicharacter classification from the Abelian case to this case. We find that if we are *very* careful, then everything carries over perfectly from the Abelian case, except for one curious but decisive twist coming from the non–commutativity of the group. The celebrated quantum *Yang–Baxter* equations rather unexpectedly appear as a consequence of the usual bicharacter relations. As we shall indicate in the final section, the rôle they play is crucial and in a certain sense they seem to determine the cocycle and ergodic action completely, which surely must come as something of a surprise. It would not have been suspected three years ago that the computation of $H^2(\hat{G}, \mathbb{T})$ would turn out to be more or less equivalent to solving the so–called 'constant symmetric quantum Yang–Baxter equations', which are currently being studied by the Russian school of Fadeev, Sklyanin and, more particularly, Drinfeld, Gurevich and Lyubashenko; and that the key to understanding this link would come from the theory of subfactors (described in the penultimate section) and the study of certain representations of S_∞.

It will be convenient to give a general definition of cocycles of $\hat{G}$ and the corresponding cocycle representations before explaining how they figure in the classification of full multiplicity ergodic actions. A *cocycle* of $\hat{G}$ is a unitary $\omega \in \mathcal{R}(G) \otimes \mathcal{R}(G)$ satisfying the cocycle identity: $\delta_G \otimes \iota(\omega)\omega_{12} = \iota \otimes \delta_G(\omega)\omega_{23}$. Two cocycles ω,ω' are called *cohomologous* or *equivalent* provided there is a unitary $v \in \mathcal{R}(G)$ such that $\omega' = \omega^v \equiv \delta_G(v^*)\omega(v \otimes v)$ and the set of equivalence classes is denoted by $H^2(\hat{G}, \mathbb{T})$. It is easily verified in the case of an abelian group that this is just the usual definition of 2–cohomology of $\hat{G}$: one uses the identification between $\mathcal{R}(G)$ and $\ell^\infty(\hat{G})$. Now let $\mathcal{M}$ be a von Neumann algebra. A *cocycle representation* of $\hat{G}$ in $\mathcal{M}$ is a unitary $W \in \mathcal{M} \otimes \mathcal{R}(G)$ such that

$f(W) = \iota \otimes \delta_G(W^*)W_{12}W_{13}$ lies in $\mathcal{R}(G) \otimes \mathcal{R}(G)$. Thus $f(W) = I \otimes \omega$ where ω is a cocycle of $\hat{G}$ and we say that W is an ω–representation. Any such W defines a coaction δ of G on $\mathcal{M}$ via the formula $\delta(x) = W(x \otimes I)W^*$ and we shall say that δ is *implemented* by W. In case $\mathcal{M} = \mathcal{B}(\mathcal{H})$, we shall refer to W as an ω–representation of $\hat{G}$ on $\mathcal{H}$. It turns out that every coaction of G on $\mathcal{B}(\mathcal{H})$ is implemented by a cocycle representation of $\hat{G}$ on $\mathcal{H}$. Now suppose that in addition to the operator $W \in \mathcal{M} \otimes \mathcal{R}(G)$ we have an action α_g of G on $\mathcal{M}$ such that $\alpha_g \otimes \iota(W) = (I \otimes \rho(g)^{-1})W$ for all $g \in G$: we say that W is an *equivariant* ω–representation of $\hat{G}$. Let us establish the existence of such representations by defining the *regular* ω–representation on $L^2(G)$ via $W_\omega = W_G \sigma\omega$ where $W_G \in L^\infty(G) \otimes \mathcal{R}(G)$ is defined by $W_G(g) = \rho(g)$. The operator W_G is fundamental in the theory of coactions. Its well known properties, together with the easily verified fact that $\sigma\omega(= \omega_{21})$ also defines a cocycle *dual* to ω, show that W_ω defines an ω–representation, equivariant with respect to the action $Ad\lambda$ of G on $\mathcal{B}(L^2(G))$. The theory of ω–representations will be clarified below once we have constructed a C* algebra universal for ω–representations of $\hat{G}$. This will immediately permit us to speak of the usual notions from representation theory such as commutants, equivalence, intertwining operators, etc. These notions can, however, be studied directly without reference to the C* algebra. For example if W is an ω–representation on $\mathcal{H}$, then the commutant of W is just the von Neumann algebra consisting of all operators T on $\mathcal{H}$ satisfying the commutation relation $[W, T \otimes I] = 0$. The bicommutant is then the commutant of this algebra and in the case of the regular ω–representation will be suggestively denoted by $\pi_\omega(\hat{G})''$. The fact that W_ω is equivariant means that G acts on the von Neumann algebra $\pi_\omega(\hat{G})''$ and this action is ergodic: the action of G is given by the restriction of $Ad\lambda$. It turns out that the commutant of $\pi_\omega(\hat{G})''$ in the $\mathcal{B}(L^2(G))$ is just $\pi_{\sigma\omega}(\hat{G})''$: this explains why we referred to $\sigma\omega$ as the cocycle 'dual' to ω. A cocycle is self–dual (or symmetric) if and only if it is trivial. Interesting as these observations are, for the purposes of ergodic actions there is a far more significant way in which ω and $\sigma\omega$ are related. One can show that the crossed product $\mathcal{M} \rtimes \hat{G}$ by a coaction implemented by an ω–representation is isomorphic to $\mathcal{M} \otimes \pi_{\sigma\omega}(\hat{G})''$ where the dual action of G is given by $\iota \otimes Ad\lambda$. This result allows one to prove the following classification theorem for ergodic actions of full multiplicity.

Classification Theorem **All ergodic actions of full multiplicity arise from the natural actions of G on $\pi_\omega(\hat{G})''$. Two ergodic actions are equivalent if and only if the correponding cocycles are cohomologous, so that there is a natural bijection between $H^2(\hat{G}, \mathbb{T})$ and ergodic actions of full multiplicity.**

To obtain the explicit link between ergodic actions and cocyles, we note that if $\alpha: G \to Aut(\mathcal{M})$ is an ergodic action of full multiplicity, then we may choose an unitary eigenmatrix U_π for each $\pi \in \hat{G}$. Defining U by $\overline{\bigoplus}_{\pi \in \hat{G}} U_\pi$ as an element in $\mathcal{M} \otimes \mathcal{R}(G)$, we obtain a unitary U such that $\alpha_g \otimes \iota(U) = U(I \otimes \rho(g))$ for all g. The ergodicity of α then implies that $W = U^*$ is an equivariant cocycle representation of $\hat{G}$ in $\mathcal{M}$, so that all the theory we have developed above becomes applicable.

Many of the facts so far outlined can be understood much better after the introduction of the L^1 and C^* algebras associated with the cocyles. In order to give clean definitions of these algebras, however, it is necessary to say a few words about *normalisation* of cocycles. Let us first recall that in the Abelian case a cocycle ω is said to be normalised if $\omega(\xi, \xi^{-1}) = 1$ for all $\xi \in \hat{G}$; and that an ω-representation U_ξ is normalised provided that $U_\xi^{-1} = U_{\xi^{-1}}$ and $U_1 = I$. In the non-Abelian case one must find substitutes for these conditions. Thus we say that a cocycle ω is *normalised* provided that $\delta_G(e_1)\omega = \delta_G(e_1)$ where e_1 is the minimal central projection in $\mathcal{R}(G)$ corresponding to the trivial representation of G; and a cocycle representation is *normalised* provided that $\iota \otimes \alpha(W) = W^*$ and $W(I \otimes e_1) = I \otimes e_1$, where α is the antipodal *-antiautomorphism $\rho(g) \mapsto \rho(g)^{-1}$ of $\mathcal{R}(G)$. We then have the following result.

Normalisation Theorem

1. **Every cocycle is equivalent to a normalised cocycle.**
2. **A cocycle is normalised if and only if the corresponding regular cocycle representation is normalised, or equivalently if any (and hence all) of its cocycle representations is normalised.**

In the Abelian case, for a normalised cocycle, one has the additional relations $\omega(\xi, 1) = 1 = \omega(1, \xi)$ together with the alternating condition $\overline{\omega(\xi, \eta)} = \omega(\xi^{-1}, \eta^{-1})$. The non-commutative analogues for a normalised cocycle are that $\omega(e_1 \otimes I) = e_1 \otimes I$, $\omega(I \otimes e_1) = I \otimes e_1$ and $\sigma\omega = \alpha\omega^*$ (where the antipode is applied to both factors). Moreover ω^v will also be normalised if and only if v satisfies the normalisation conditions $\alpha v = v^*$ and $ve_1 = e_1$. To prove all these assertions, one has to make constant use of the elementary but nevertheless fundamental properties of the 'diagonal' element $\Delta = \delta_G(e_1)$ of $\mathcal{R}(G) \otimes \mathcal{R}(G)$:

1. $\Delta(I \otimes x) = \Delta(\alpha x \otimes I)$.
2. $\Delta(I \otimes x) = 0$ if and only if $x = 0$.
3. $\Delta(\mathcal{R}(G) \otimes \mathcal{R}(G)) = \Delta(I \otimes \mathcal{R}(G))$.

We are now in a position to define the algebras $L^1_\omega(\hat{G})$ and $C^*_\omega(\hat{G})$, where ω is a normalised cocycle of $\hat{G}$. We define $L^1_\omega(\hat{G})$ to be $\mathcal{R}(G)_*$, the predual of $\mathcal{R}(G)$, with its usual Banach space structure and involution but with multiplication perturbed by ω. Thus $\langle \phi \circ \psi, x \rangle = \langle \phi \otimes \psi, \delta_G(x)\omega \rangle$ for $\phi, \psi \in \mathcal{R}(G)_*$ and $x \in \mathcal{R}(G)$. This makes $L^1_\omega(\hat{G})$ into a unital Banach *-algebra with its usual unit ε, where ε is the trivial character. (We recall that when ω is the trivial cocycle, we just recover the usual Fourier algebra of G, which may be identified with an algebra of functions on G using the map $\phi \mapsto \hat{\phi}$ where $\hat{\phi}(g) = \phi(\rho(g))$.) One checks that the formula $\pi_W(\phi) = \iota \otimes \phi(W)$ establishes a one-one correspondence between ω-representations of $\hat{G}$ and unital *-representations of $L^1_\omega(\hat{G})$. In the course of establishing this it is important to consider the special case of the regular ω-representation. Here one starts by noting that G acts on the algebra $L^1_\omega(\hat{G})$ according to the formula $\alpha_g(\phi) = \phi \cdot \rho(g^{-1})$ and that, since it coincides with the action by left translation on functions, this action is ergodic. The unique G-invariant state on $L^1_\omega(\hat{G})$

given by averaging is therefore just $\tau(\phi) = \phi(e_1)$. Its associated GNS–representation π_τ is equivariantly equivalent to the regular representation π_{W_ω}. This dictionary between ω–representations and unital *–representations of $L^1_\omega(\hat{G})$ can of course be extended to cover general equivariant representations, commutants, intertwining operators, and so forth. There are no surprises. We can then define $C^*_\omega(\hat{G})$ to be the enveloping C* algebra of $L^1_\omega(\hat{G})$. The action of G on $L^1_\omega(\hat{G})$ extends to an ergodic action of G on $C^*_\omega(\hat{G})$ of full multiplicity. It is a nuclear C* algebra, since $C^*_\omega(\hat{G}) \rtimes G \cong \mathcal{K}$ is nuclear, and can be identified with the C* algebra generated by the spectral subspaces of $\pi_\omega(\hat{G})''$. As we shall see its spectrum always has the form G/H as a G–space, so that $C^*_\omega(\hat{G})$ is simple if and only if $\pi_\omega(\hat{G})''$ is a factor. The justification for this lies in the observation of Magnus Landstad that if a compact group acts on a primitive C* algebra, then every non–zero ideal in the algebra contains a non–zero invariant ideal. Landstad's proof of this was extracted from some ingenious computations of Olesen and Pedersen. Let us give here an alternative more conceptual proof. Any non–zero ideal corresponds to a closed subset of the primitive ideal space, missing the zero ideal. By a result of Glimm, the compact group acts continuously on the primitive ideal space so that the saturation of the closed set is automatically closed. It is also by definition invariant and misses the zero ideal. The corresponding ideal therefore provides the required non–zero invariant ideal.

The question naturally arises as to when $\pi_\omega(\hat{G})''$ can be a factor. In the Abelian case, this question can be answered by introducing the bicharacter associated to a cocycle. Let us briefly indicate how this is effected. Firstly when G is Abelian, there is an isomorphism of $H^2(\hat{G}, \mathbb{T})$ into the group of alternating bicharacters on $\hat{G}$ given by $[\omega] \mapsto \beta$ where $\beta(\xi,\eta) = \omega(\xi,\eta)\overline{\omega(\eta,\xi)}$. Thus β is separately multiplicative in each variable and satisfies the alternating condition $\overline{\beta(\xi,\eta)} = \beta(\eta,\xi)$. The bicharacter is said to be *non–degenerate* or *totally skew* if the usual condition holds: $\beta(\xi,\eta) = 1$ for all η if and only if $\xi = 1$. It is also usual to interpret this condition in terms of the homomorphism Λ of $\hat{G}$ into G defined by $\Lambda(\xi) = \beta(\xi,\cdot)$. One obtains the following criteria for $\pi_\omega(\hat{G})''$ to be a factor:

Factoriality Theorem **If ω is a cocycle of $\hat{G}$ with corresponding bicharacter β, the following conditions are equivalent:**

1. **$\pi_\omega(\hat{G})''$ is a factor.**
2. **Λ is injective.**
3. **Λ has dense image.**
4. **β is non–degenerate.**
5. **$C^*_\omega(\hat{G})$ (or $L^1_\omega(\hat{G})$) has a unique trace.**
6. **$C^*_\omega(\hat{G})$ (or $L^1_\omega(\hat{G})$) has trivial centre.**
7. **$C^*_\omega(\hat{G})$ is primitive.**
8. **$C^*_\omega(\hat{G})$ is simple.**

These results are all fairly straightforward to prove in the Abelian case, and our task now will be to show that by appropriately defining bicharacters β and associated maps Λ everything carries over to the non–Abelian setting. As we have already hinted, however, the non–commutativity of the group will nevertheless give rise to a fundamental and crucial point of departure from the Abelian prototype. Before giving the analogous definitions, let us pause to explain Landstad's method of establishing the last condition from the others. The customary method proceeded by observing that the subgroup $\Lambda(\hat{G})$ of G always acts by inner automorphisms on $C^*_\omega(\hat{G})$. Thus when Λ is dense, the action of G has to be approximately inner and therefore any ideal of $C^*_\omega(\hat{G})$ is automatically G–invariant. Since the crossed product is simple, the original algebra has to be simple. All we needed to make the last part of the argument work was *some* non-trivial G–invariant ideal in $C^*_\omega(\hat{G})$: this could have been produced equally easily, however, by assuming $C^*_\omega(\hat{G})$ not to be simple and then using Landstad's observation. This reasoning of course has the advantage of extending to the case of actions of non–commutative groups.

We are now ready to introduce the formalism of bicharacters. This is closely related to Drinfeld's use of "triangular Hopf algebras" for studying the quantum Yang–Baxter equations although we have avoided using the language of Kac (or Hopf–von Neumann) algebras too much. (We shall give a fuller explanation for the link with the work of Drinfeld and his school in Section VII.) If ω is a normalised cocycle of $\hat{G}$, we define the *bicharacter* β_ω of ω to be the unitary $\beta_\omega = (\sigma\omega^*)\omega = \omega_{21}^*\omega_{12}$ in $\mathcal{R}(G) \otimes \mathcal{R}(G)$. Thus if W is an ω–representation, then $\beta_\omega = W_{12}^* W_{13}^* W_{12} W_{13}$. We define a new comultiplication δ_ω on $\mathcal{R}(G)$ by $\delta_\omega(x) = \omega^*\delta_G(x)\omega$ in order to reveal the bicharacter nature of β_ω. This induces a new unital Banach *–algebra structure on the predual $\mathcal{R}(G)_*$, which we shall denote by $A_\omega(G)$ when it has this structure. The perturbed comultiplication now satisfies $\sigma\delta_\omega(x) = \beta\delta_\omega(x)\beta^*$, so that the multiplication on $A_\omega(G)$ is no longer commutative, unlike the case of the usual Fourier algebra $A(G)$; it is evident, however, that the multiplication on the space of class functions remains unaltered so that it forms a commutative subalgebra of $A_\omega(G)$. It will not in general be central. One can also show that any faithful set of matrix coefficients will generate $A_\omega(G)$ as a *–algebra. We summarise the properties of the pair $(\beta_\omega, \delta_\omega)$ (which we shall simply write as (β, δ)):

1. (Alternating) $\beta^* = \sigma\beta$, $\alpha\beta = \beta$, $\iota \otimes \alpha(\beta) = \beta^* = \alpha \otimes \iota(\beta)$
2. (Normalised) $(e_1 \otimes I)\beta = e_1 \otimes I$, $(I \otimes e_1)\beta = I \otimes e_1$, $\delta(e_1)\beta = \delta(e_1)(= \Delta)$
3. (Cocycle relations) $\beta_{12}\delta \otimes \iota(\beta) = \beta_{23}\iota \otimes \delta(\beta)$
4. (Bicharacter relations) $\delta \otimes \iota(\beta) = \beta_{13}\beta_{23}$, $\iota \otimes \delta(\beta) = \beta_{13}\beta_{12}$
5. (Quantum Yang–Baxter equations) $\beta_{12}\beta_{13}\beta_{23} = \beta_{23}\beta_{13}\beta_{12}$

Two pairs (β_1, δ_1) and (β_2, δ_2) are said to be *equivalent* if there is a normalised unitary $v \in \mathcal{R}(G)$ such that $\beta_1 = (v^* \otimes v^*)\beta_2(v \otimes v)$ and $\delta_1 = Ad(v^* \otimes v^*) \cdot \delta_2 \cdot Ad(v)$. This is consistent with the notion of equivalence on cocycles, although if G is not connected the notions need not be equivalent. In general the equivalence relation on pairs when pulled back to $H^2(\hat{G}, \mathbb{T})$ via the map $\omega \mapsto (\beta_\omega, \delta_\omega)$ induces an extra equivalence relation. To explain this weaker equivalence relation we must recall a definition due to Burnside. Let $Aut_c(G)$ be the group of automorphisms of G acting trivially on $\hat{G}$. Such automorphisms can equivalently be described by the triviality of their action on class functions or by the fact that they may be implemented by conjugation by a unitary in $\mathcal{R}(G)$. This normal subgroup of $Aut(G)$ clearly contains the group $Inn(G)$ of inner automorphisms. The quotient $Aut_c(G)/Inn(G)$ is trivial if G is connected and otherwise is a projective limit of finite solvable groups (by work of Sah). Now for an element $\gamma \in Aut_c(G)$, let $u \in \mathcal{R}(G)$ be a (normalised) unitary such that $Ad(u)\rho(g) = \rho(\gamma(g))$ for all $g \in G$. Then the map $\omega \mapsto Ad(u \otimes u)\omega$ defines an action of $Aut_c(G)/Inn(G)$ on $H^2(\hat{G}, \mathrm{T})$. The main result is that two pairs are equivalent if and only if the corresponding cocycles lie in the same orbit of $Aut_c(G)/Inn(G)$ in $H^2(\hat{G}, \mathrm{T})$.

Let us now define the remaining ingredients required for the Factoriality Theorem in the non–commutative case. We shall say that β is *non–degenerate* provided that $(x \otimes I)\beta = x \otimes I$ implies that x is a scalar multiple of e_1. The rôle of the map $\Lambda: \hat{G} \to G$ is played (dually) by a *–homomorphism $\Lambda: A_\omega(G) \to \mathcal{R}(G)$ defined by $\Lambda(\phi) = (\phi \otimes \iota)\beta$. The theorem then holds without change (except that the image of Λ should now be dense in the ultraweak topology on $\mathcal{R}(G)$). One has to use the new tensor product on representations of G defined by the perturbed comultiplication δ_ω; the formulas for decomposing tensor products remain the same as before, but the operator β must now be used to exhibit the isomorphism between $V \otimes W$ and $W \otimes V$. The map Λ appears in the picture because the ultraweak closure of its image $\Lambda(A_\omega(G)''$ is a Kac subalgebra of $\mathcal{R}(G)$: it is always of the form $v^*\mathcal{R}(H)v$ for some closed subgroup H of G and the primitive spectrum of $C^*_\omega(\hat{G})$ is then just G/H. The only tricky part of the theorem lies in finding a purely Hopf–algebraic proof of the assertion that if the restriction of every irreducible representation of $\mathcal{R}(G)$ to some Kac subalgebra never contains the trivial representation, then the subalgebra must be the whole of $\mathcal{R}(G)$. Of course the 'classical' proof would use Frobenius Reciprocity.

We shall close this section by showing how the bicharacters can be classified for four examples, namely $SU(2)$, $SO(3)$, $SU(2) \times SU(2)$ and $SU(3)$, although the techniques are applicable to other groups such as $U(2)$, $PU(3)$, $SO(4)$, etc. In these cases, however, it is often easier to work with a covering group and use the theory of the multiplicity matrix.

Example 1: $SU(2)$. We will prove that if $\alpha: SU(2) \to Aut(\mathcal{M})$ is an ergodic action in which at least one copy of the two–dimensional irreducible representation π of $SU(2)$ occurs, then $\mathcal{M} \cong L^\infty(SU(2))$ as an $SU(2)$–algebra. This will show that $SU(2)$ has no non–trivial bicharacters. So suppose that we can find non–zero elements a and b in $\mathcal{M}$ such that

$$\alpha_g(a) = \alpha a - \overline{\beta} b$$
$$\alpha_g(b) = \beta a + \overline{\alpha} b,$$

for all $g \equiv \pi(g) = \begin{pmatrix} \alpha & \beta \\ -\overline{\beta} & \overline{\alpha} \end{pmatrix}$ in $SU(2)$. Let $M = \begin{pmatrix} a & b \\ -b^* & a^* \end{pmatrix}$. Then M is an eigenmatrix for π, in that $\alpha_g(M) = M\pi(g)$ for all $g \in SU(2)$. Hence $\alpha_g(MM^*) = MM^*$, so that the entries of MM^* are scalars by the ergodicity of α. Thus $aa^* + bb^*$, $b^*b + a^*a$ and $ab - ba$ are scalars. Taking traces and normalising so that $aa^* + bb^* = 1$, we find that $MM^* = I$. Since $\mathcal{M}$ is finite, this forces the M to be unitary, i.e. $M^*M = I$. Hence

$a^*a + bb^* = 1$ and $a^*b = ba^*$. Therefore a and b generate an Abelian *– algebra which is $SU(2)$–invariant. We can then either use Frobenius Reciprocity or our general eigenmatrix argument to deduce that the ultraweak closure of this algebra is isomorphic to $L^\infty(SU(2))$ as an $SU(2)$–algebra. Since every representation of $SU(2)$ must therefore occur with full multiplicity in the algebra generated by a and b, it follows that a and b generate the whole of $\mathcal{M}$ and our claim is established.

Example 2: $SO(3)$. We shall next establish that $SO(3)$ has just two bicharacters, the trivial one and the one induced from the non–trivial bicharacter of the Klein 4–group. We shall do this by showing that any ergodic action of $SO(3)$ of full multiplicity is necessarily on a Type I algebra; this is clearly equivalent to the statement on bicharacters. So suppose that $\alpha: SO(3) \to Aut(\mathcal{M})$ is an ergodic action of full multiplicity. By choosing self–adjoint orthogonal bases a_i, b_i, c_i for the three copies of the three–dimensional representation π_1, we can produce a unitary eigenmatrix $M = \begin{pmatrix} a_1 & b_1 & c_1 \\ a_2 & b_2 & c_2 \\ a_3 & b_3 & c_3 \end{pmatrix}$ for π_1. We next introduce the non–commutative determinants μ_{ijk} defined by

$$\begin{aligned} \mu_{ijk} &=: \det: \begin{pmatrix} a_i & b_i & c_i \\ a_j & b_j & c_j \\ a_k & b_k & c_k \end{pmatrix} \\ &= (a_ib_j - b_ia_j)c_k + (b_ic_j - c_ib_j)a_k + (c_ia_j - a_ic_j)b_k \\ &= 6Tr(a_ib_jc_k). \end{aligned}$$

That these determinants are scalars follows from two easy facts: namely, if a, b, c and a', b', c' are two copies of π_1 in an $SO(3)$–algebra, then on the one hand $aa' + bb' + cc'$ lies in the fixed point algebra; while on the other hand, the *wedge products* $a'' = bc' - cb'$, $b'' = ca' - ac'$, $c'' = ab' - ba'$ give another copy of π_1. It follows immediately from the properties of the trace, that the determinant μ_{ijk} is invariant under cyclic permutations of the indices (which need not be distinct) and also satisfies $\overline{\mu_{ijk}} = -\mu_{kji}$. To establish further algebraic properties of these determinants, we introduce the matrices A_i and B_{ij} defined as follows

$$A_i = \begin{pmatrix} 0 & -c_i & b_i \\ c_i & 0 & -a_i \\ -b_i & a_i & 0 \end{pmatrix} \quad \text{and} \quad B_{ij} = \begin{pmatrix} a_ia_j & b_ia_j & c_ia_j \\ a_ib_j & b_ib_j & c_ib_j \\ a_ic_j & b_ic_j & c_ic_j \end{pmatrix}.$$

We note that if X is any of these matrices, then $\alpha_g(X) = \pi_1(g)^* X \pi_1(g)$ for $g \in SO(3)$ and hence MXM^* has scalar entries. In particular the (i,j) entry of MA_kM^* is just μ_{ikj}. Let us list some of the properties of the matrices A_i and B_{ij}. By definition A_i is skew–adjoint with trace zero. Moreover $B_{ij} = A_iA_j + \delta_{ij}I$, $\sum_i A_i^2 = -2I$ and $tr(A_iA_j) = -2\delta_{ij}$.

Now the skew property $\overline{\mu_{ijk}} = -\mu_{kji}$ and cyclic property $\mu_{ijk} = \mu_{jki}$ of the determinant μ_{ijk} show that it has the form

$$\mu_{ijk} = \sigma\varepsilon_{ijk} + i\rho_{ijk}$$

where σ is real, ε_{ijk} is the alternating tensor and ρ_{ijk} is (completely) symmetric and real. Note that $\rho_{ijk} = Im(6tr(a_i b_j c_k))$. We claim that we may choose the bases a_i, b_i, c_i so that $\rho_{111} = 0 = \rho_{211}$. Now in any subspace formed by two copies of π_1, we can find a basis a, b, c satisfying $Im(tr(abc) = 0$, since this just amounts to finding a real root of a cubic with real coefficients. So we can choose a_1, b_1, c_1 so that $\rho_{111} = 0$. The wedge products of this basis with itself are then orthogonal to it. If they are zero, there is nothing to prove. Otherwise, we take the second basis a_2, b_2, c_2 to be orthogonal to the first basis and its wedge product with itself. We now write out the matrices $(\rho_{ijk})_{i,k}$ explicitly imposing all the conditions mentioned above. We find that they become

$$\begin{pmatrix} 0 & 0 & e \\ 0 & b & f \\ e & f & -b \end{pmatrix}, \quad \begin{pmatrix} 0 & b & f \\ b & y & z \\ f & z & -y \end{pmatrix}, \quad \begin{pmatrix} e & f & -b \\ f & z & -y \\ -b & -y & -e-z \end{pmatrix}$$

where $eb = 0$ and $1 - \sigma^2 = e^2 + b^2 + f^2 = b^2 + f^2 + y^2 + z^2 = e^2 + z^2 + f^2 + ez + y^2 + b^2$. These relations imply that $z = -e$ and $y = 0$ and that either e or b is zero. We have just one relation remaining after these deductions, namely $\sigma^2 + e^2 + f^2 + b^2 = 1$. Now if $e = 0$, then we find that

$$M(a_1 \otimes I)M^* = \begin{pmatrix} 1 & 0 & 0 \\ 0 & \lambda & 2\sigma ib \\ 0 & 2\sigma ib & \lambda^* \end{pmatrix} a_1 \otimes I$$

where $\lambda = -b^2 + (if - \sigma)^2$. (This follows by rewriting MB_{1k} in the form $(MB_{1k}M^*)M$.) The self–adjointness of a_1 implies that $\sigma b = \sigma f = 0$ so that $M(a_1^2 \otimes I)M^* = a_1^2 \otimes I$. Thus a_1^2, and hence all products from the first basis, is central. Similarly if $b = 0$, we see that all products of elements in the third basis a_3, b_3, c_3 are central. Now an easy extension of the eigenmatrix duality criterion allows us to deduce that the (central) algebra generated by either of these sets of products is equivariantly isomorphic to $L^\infty(SO(3)/N(T))$, where $N(T)$ is the normaliser of a maximal torus, i.e. the infinite dihedral group D_∞^*. Thus the ergodic action of $SO(3)$ must be induced from an ergodic action of $N(T)$, and hence the bicharacters of $\widehat{SO(3)}$ are as claimed.

Example 3: $SU(2) \times SU(2)$. We will show that all bicharacters of $SU(2) \times SU(2)$ are induced from bicharacters of a maximal torus and therefore are classified by a parameter in the circle group. So suppose that we have a full multiplicity ergodic action of $G = SU(2) \times SU(2)$ on $\mathcal{M}$. Let the two factors of G be G_1 and G_2. Then we have full multiplicity ergodic actions of G_1 and G_2 on $\mathcal{M}^{G_2}$ and $\mathcal{M}^{G_1}$ respectively. By the result of Example 1, for each factor G_i we can find unitary eigenmatrices $M_i = \begin{pmatrix} a_i & b_i \\ -b_i^* & a_i^* \end{pmatrix}$ lying in the fixed point algebra of the other factor. Each M_i is unique up to multiplication by an element of $SU(2)$. By uniqueness it follows that $M_1(a_2 \otimes I)M_1^* = Pa_2 - Qb_2^*$ where P and Q are scalar matrices. The unitarity of M_2 then implies that the matrix $\begin{pmatrix} P & Q \\ -Q^* & P^* \end{pmatrix}$ is unitary, so that P and Q are in particular normal and commute. Premultiplying M_1 by an element of $SU(2)$ has the effect of conjugating P and Q by that same element. We may therefore assume that P and Q are in diagonal form and a simple manipulation shows that their diagonal entries are complex conjugates so that $P = \begin{pmatrix} p & 0 \\ 0 & \overline{p} \end{pmatrix}$ and $Q = \begin{pmatrix} q & 0 \\ 0 & \overline{q} \end{pmatrix}$ with $|p|^2 + |q|^2 = 1$. The final step consists of checking what happens if we adjust the eigenmatrix M_2 by premultiplying by some element $\begin{pmatrix} \alpha & \beta \\ -\overline{\beta} & \overline{\alpha} \end{pmatrix}$ of $SU(2)$. It turns out that we just have to conjugate $\begin{pmatrix} p & q \\ -q^* & p^* \end{pmatrix}$ by this element. Thus we reach a final canonical form by conjugating by an element so that $q = 0$. The Riedel rotation $p \in \mathbf{T}$ is thus the only remaining invariant. Now consider the *–algebra generated by the products xy^* where $x, y \in \{a_i, b_i\}$ $(i = 1, 2)$. It is Abelian because the canonical form shows that these products commute with both M_1 and M_2. As in the previous example, we may use an argument similar to that employed in the proof of the eigenmatrix criterion for duality to show that this algebra is equivariantly isomorphic to $L^\infty(G/T)$ where T is a maximal torus in G. Thus the ergodic action must be induced from a maximal torus, as claimed.

Example 4: $SU(3)$. We will sketch a proof that all bicharacters (or equivalently full multiplicity ergodic actions) of $SU(3)$ are induced form those of a maximal torus. Actually we will content ourselves with proving the slightly weaker statement that any algebra admitting a full multiplicity ergodic action of $SU(3)$ has a non–trivial (and fairly large) centre. Our proof has two ingredients: namely the formalism of determinants, generalising the techniques of Example 2; and the existence of points of inflexion for plane cubic curves defined over $\mathbf{C}$. We already implicitly considered a real cubic cubic curve when working with $SO(3)$, namely the curve defined by $\mu_{111} = 0$. To be more precise, this was the curve $F(\lambda) \equiv tr(a_\lambda b_\lambda c_\lambda) = 0$ where $x_\lambda = \lambda_1 x_1 + \lambda_2 x_2 + \lambda_3 x_3$ in the real projective plane with coordinates $(\lambda_1 : \lambda_2 : \lambda_3)$.

Let us start by introducing the appropriate versions of the matrices A_i and B_{ij}. Let $M = \begin{pmatrix} a_1 & b_1 & c_1 \\ a_2 & b_2 & c_2 \\ a_3 & b_3 & c_3 \end{pmatrix}$ be a unitary eigenmatrix for the three–dimensional (identity) representation π of $SU(3)$, so that $\alpha_g(M) = M\pi(g)$ for all $g \in SU(3)$. The entries of M are of course no longer self–adjoint and M is now unique up to multplication by a unitary, rather than an orthogonal, matrix. We take the same definitions for the determinants μ_{ijk} and matrices A_i as before. This time, however, the representation π is no longer self–conjugate and therefore the definition of the B_{ij}'s must be modified. It is still true, however, that the matrix obtained by applying the involution to the entries of M is a unitary eigenmatrix for the conjugate representation $\overline{\pi}$. As a consequence the matrix M^t will also be unitary and satisfy $\alpha_g(M^t) = \pi(g)^t M^t$. We define

$$B_{ij} = \begin{pmatrix} a_i a_j^* & b_i a_j^* & c_i a_j^* \\ a_i b_j^* & b_i b_j^* & c_i b_j^* \\ a_i c_j^* & b_i c_j^* & c_i c_j^* \end{pmatrix}.$$

We then find that $\alpha_g(A_i) = \pi(g)^* A_i(\pi(g)^t)^*$ and $\alpha_g(B_{ij}) = \pi(g)^* B_{ij}\pi(g)$, so that MA_iM^t and $MB_{ij}M^*$ are scalar matrices. (Note that analogous matrices to the B_{ij} may be defined for unitary eigenmatrices corresponding to an (irreducible) representation π of any compact Lie group.) The corresponding scalar tensor $\beta_{ijkl} = (MB_{ij}M^*)_{kl}$ can easily be identified with the (π, π) component of the corresponding bicharacter and in particular satisfies the quantum Yang–Baxter equation. As before the (i, k) entry of the matrix MA_jM^t is the determinant $\mu_{ijk} = 6tr(a_ib_jc_k)$. Furthermore we have the relations $B_{ij} = \delta_{ij}I - A_iA_j^*$, $\sum_i B_{ii} = I$, $tr(B_{ij}) = \delta_{ij}$, $tr(B_{ii}^2) = 1$, and $(M(a_k \otimes I)M^*)_{ij} = \sum_\ell \beta_{kji\ell}a_\ell$. We also have the symmetry condition for the Yang–Baxter equation, which may be stated in terms of the operator $\Delta = \begin{pmatrix} B_{11} & B_{12} & B_{13} \\ B_{21} & B_{22} & B_{23} \\ B_{31} & B_{32} & B_{33} \end{pmatrix}$. Thus we require that Δ be a self–adjoint unitary. We shall only use the fact that the B_{ij}'s arise from the determinants μ_{pqr}. These may be regarded as providing a 'square root' of the Yang–Baxter solution. In general we can modify μ by replacing M by AM where $A \in U(3)$. By definition the tensor μ_{ijk} is *cyclic*, i.e. $\mu_{ijk} = \mu_{jki}$, and hence it may be written uniquely as the sum of an alternating tensor α and a symmetric tensor σ

$$\mu_{ijk} = \sigma_{ijk} + \alpha_{ijk}$$

where $\sigma_{ijk} = \mu_{ijk}$ unless $\{i, j, k\} = \{1, 2, 3\}$ when it is given by $\frac{1}{2}(\mu_{123} + \mu_{213})$; and where $\alpha_{ijk} = \frac{1}{2}(\mu_{123} - \mu_{213})\varepsilon_{ijk}$. Let us consider the cubic curve in $\mathbb{C}P^2$ defined by $\mu_{111} = 0$ as A varies. Thus if $(\alpha_1, \alpha_2, \alpha_3)$ is the first row of A, we are considering the curve whose homogeneous coordinates $(\alpha_1{:}\,\alpha_2{:}\,\alpha_3)$ satisfy $F(\alpha) = tr(a_\alpha b_\alpha c_\alpha) = 0$, where $x_\alpha = \alpha_1x_1 + \alpha_2x_2 + \alpha_3x_3$. The equation of the cubic may be rewritten as $\sum \mu_{ijk}\alpha_i\alpha_j\alpha_k = 0$ or yet again as $\sum \sigma_{ijk}\alpha_i\alpha_j\alpha_k = 0$. Exactly as for $SO(3)$ we may arrange for $\mu_{111} = 0 = \mu_{112}$; however, if we demand in addition that $\mu_{113} = 0$, then this is equivalent to the cubic having a singular point which we cannot be sure happens. So let us suppose first that the contrary holds and our cubic is non–singular. Now a non–singular cubic is known to have exactly nine points of inflection. Since $U(3)$ acts transitively on the sphere bundle of $\mathbb{C}P^2$, we can always arrange for $(0{:}\,1{:}\,0)$ to be the point of inflection and $\alpha_3 = 0$ to be the tangent there. Thus we can find A so that $\sigma_{112} = \sigma_{221} = \sigma_{222} = 0$. Thus $F(\alpha)$ has the form $F(\alpha) = \sigma_{333}\alpha_3^3 + \sigma_{111}\alpha_1^3 + 3\sigma_{332}\alpha_3^2\alpha_2 + 3\sigma_{311}\alpha_3\alpha_1^2 + 3\sigma_{223}\alpha_3\alpha_2^2 + 3\sigma_{331}\alpha_3^2\alpha_1 + \sigma_{123}\alpha_1\alpha_2\alpha_3$. Let $C_i = MA_iM^t$. These matrices have the form

$$C_1 = \begin{pmatrix} a & 0 & p \\ 0 & 0 & f \\ p & g & q \end{pmatrix}, \quad C_2 = \begin{pmatrix} 0 & 0 & g \\ 0 & 0 & s \\ f & s & \ell \end{pmatrix}, \quad C_3 = \begin{pmatrix} p & f & q \\ g & s & \ell \\ q & \ell & r \end{pmatrix}.$$

Let $X = C_2C_2^*$. Then $\det(X) = 0$ and $Tr(X) = Tr(X^2) = 2$. Thus X has to be a projection. Similarly $Y = C_2^*C_2$ must be a projection. Let us suppose first that $g = 0$. Then the identity $X^2 = X$ forces $|s| = 1$ and $\ell = f = 0$. Similarly if $f = 0$, then $Y^2 = Y$ implies that $|s| = 1$ and $\ell = g = 0$. X and Y must then both be equal to $\begin{pmatrix} 1 & 0 & 0 \\ 0 & 0 & 0 \\ 0 & 0 & 1 \end{pmatrix}$, so that $|a|^2 + |p|^2 = 1 = |p|^2 + |q|^2$ and $a\overline{p} + p\overline{q} = 0$. Using the identity $\sum_i C_i C_i^* = 2I$, it is easy to see that q, r and a must vanish and we may arrange that $p = s = 1$. Exactly as in the calculations for $SO(3)$, we can prove that $M(x \otimes I)M^* = \begin{pmatrix} -1 & 0 & 0 \\ 0 & -1 & 0 \\ 0 & 0 & 1 \end{pmatrix} (x \otimes I)$ when x is an element in the third basis. It follows that a_3, b_3 and c_3 generate an Abelian C* algebra such that all elements of the form xy or xy^* are central, where x and y are any elements of the basis. It is not too hard to see that the C* algebra generated by these products is of the form $C(SU(3)/H)$ where H is the subgroup $SU(3) \cap (U(2) \times U(1))$ of $SU(3)$. Thus at least one knows that the action is induced from a full multiplicity ergodic action of $H \cong U(2)$. (To show that it is in fact induced from an action of a maximal torus, one can make use of the extra information that the two new bases $a_1 \pm ia_2$, $b_1 \pm ib_2$, $c_1 \pm ic_2$ satisfy exactly the same conditions as the third basis a_3, b_3 and c_3. One then verifies that the central C* algebra generated by all elements of the form xy or xy^*, where x and y lie in the same basis, has the form $C(SU(3)/T)$ for some maximal torus T in $SU(3)$.)

Now we shall assume that g and f are non–zero, but the cubic is still non–singular. From the equations $X^2 = X$ and $Y^2 = Y$ we deduce that $\ell = 0$, $|f| = |g|$ and $|g|^2 + |s|^2 = 1$. We may then arrange that $s = \sin(\theta)$, $g = \cos(\theta) \neq 0$ and $f = \zeta \cos(\theta)$ where $|\zeta| = 1$. We then look at the relations implied by the identity $\sum_i C_i C_i^* = 2I$. We find in particular that $p(\overline{f} + \overline{g}) + (f + g)s = 0$. On the other hand, if $f + g = 0$, then it is easy to check that the cubic has a singular point, contrary to assumption. Thus $f + g$ cannot vanish and therefore $p = -\zeta \sin(\theta)$. The remaining relations then show that $a = q = r = 0$ so that finally we find that the C_i's have the canonical form

$$C_1 = \begin{pmatrix} 0 & 0 & -\zeta s_\theta \\ 0 & 0 & \zeta c_\theta \\ -\zeta s_\theta & c_\theta & 0 \end{pmatrix}, \quad C_2 = \begin{pmatrix} 0 & 0 & c_\theta \\ 0 & 0 & s_\theta \\ \zeta c_\theta & s_\theta & 0 \end{pmatrix}, \quad C_3 = \begin{pmatrix} -\zeta s_\theta & \zeta c_\theta & 0 \\ c_\theta & s_\theta & 0 \\ 0 & 0 & 1 \end{pmatrix},$$

where $c_\theta = \cos(\theta)$ and $s_\theta = \sin(\theta)$. Exactly as in the first case we may show that all elements of the form xy^*, where x and y are drawn from the third basis, must be central. Indeed we find that for any element x from the third basis we have $M(x \otimes I)M^* = \begin{pmatrix} \alpha & \beta & 0 \\ -\overline{\beta} & \overline{\alpha} & 0 \\ 0 & 0 & 1 \end{pmatrix} (x \otimes I)$ where $\alpha = -s_\theta^2 - \zeta c_\theta^2$ and $\beta = c_\theta s_\theta (1 - \zeta)$. Thus we find that M commutes with all elements of the form xy^* as required. Consequently our claim is established when the cubic is non–singular.

When the cubic has a singular point, we can assume that $\mu_{111} = \mu_{112} = \mu_{113} = 0$. This is equivalent to choosing a basis a_1, b_1, c_1 whose elements commute. The matrices C_i then assume the following form:

$$C_1 = \begin{pmatrix} 0 & 0 & 0 \\ 0 & a & f \\ 0 & g & b \end{pmatrix}, \quad C_2 = \begin{pmatrix} 0 & a & g \\ a & x & c \\ f & c & y \end{pmatrix}, \quad C_3 = \begin{pmatrix} 0 & f & b \\ g & c & y \\ b & y & z \end{pmatrix}.$$

Let $X = C_1C_1^*$. Then $\det(X) = 0$ and $Tr(X) = Tr(X^2) = 2$. From this it follows that $C_1C_1^* = \begin{pmatrix} 0 & 0 & 0 \\ 0 & 1 & 0 \\ 0 & 0 & 1 \end{pmatrix}$ so that C_1 has the form $C_1 = \begin{pmatrix} 0 & 0 & 0 \\ 0 & s_\theta & c_\theta\zeta \\ 0 & -c_\theta\overline{\zeta} & s_\theta \end{pmatrix}$, where $c_\theta = \cos(\theta)$, $s_\theta = \sin(\theta)$ and $|\zeta| = 1$. Here we have used the equations $a = 6Tr(a_2b_1c_2)$ and $b = 6Tr(a_3b_1c_3)$ to make a priori homotheties of the second and third bases to ensure that $a, b \geq 0$. Now the condition $\sum_i C_iC_i^* = 2I$ immediately implies that $x = c = y = z = 0$. Thus the matrices C_i can be put in the canonical form

$$C_1 = \begin{pmatrix} 0 & 0 & 0 \\ 0 & s_\theta & c_\theta\zeta \\ 0 & -c_\theta\overline{\zeta} & s \end{pmatrix}, \quad C_2 = \begin{pmatrix} 0 & s_\theta & -c_\theta\overline{\zeta} \\ s_\theta & 0 & 0 \\ c_\theta\zeta & 0 & 0 \end{pmatrix}, \quad C_3 = \begin{pmatrix} 0 & c_\theta\zeta & s_\theta \\ -c_\theta\overline{\zeta} & 0 & 0 \\ s_\theta & 0 & 0 \end{pmatrix}.$$

By employing arguments similar to those used in the previous cases, one can then show all elements of the form xy^*, with x and y coming from the first basis, must be central in $\mathcal{M}$. Indeed for such elements one has $M(x \otimes I)M^* = \begin{pmatrix} 1 & 0 & 0 \\ 0 & \alpha & \beta \\ 0 & -\overline{\beta} & \overline{\alpha} \end{pmatrix} (x \otimes I)$ where $\alpha = \zeta^2 c_\theta^2 - s_\theta^2$ and $\beta = c_\theta s_\theta(\zeta + \overline{\zeta})$, so that $xy^* \otimes I$ commutes with M. This completes our discussion for $SU(3)$.

VI. An Invariance Principle for Subfactors

One of the main problems in the theory of subfactors (initiated by Vaughan Jones) is the determination of the two towers of higher relative commutants associated with the extensions of a factor by a subfactor obtained by iterating the Jones basic construction. These towers have certain remarkable properties which have been put into evidence by the work of Adrian Ocneanu, Mihai Pimsner, and Sorin Popa. These properties have been formalised by Ocneanu starting from the basic observation that the structure of the higher relative commutant towers of the factor and subfactor can be understood independently of the Jones basic construction by considering the natural operations of induction, restriction and conjugation between the four classes of bimodules associated with the factor and subfactor. For subfactors of finite index and finite depth (i.e. for which the Bratteli diagrams for the inclusions in the relative commutant towers eventually become periodic), Ocneanu now appears to have proved a wonderful rigidity theorem. Roughly speaking this states that the inclusion of factors obtained by taking the limiting algebras in the relative commutant towers is isomorphic to the original inclusion of factors. In particular, this leads to a complete classification of subfactors of index less than four, that is for the special values $4\cos^2(\pi/n)$. His methods, however, appear not to extend to the case of infinite depth where very little is known at present, even for integer indices. Our aim here is to show how a very simple invariance principle permits the painless determination of higher relative commutants for various subfactors of infinite depth associated with compact groups.

Let $\mathcal{N}$ be a subfactor of the hyperfinite II_1 factor $\mathcal{M}$ of finite index $[\mathcal{M}:\mathcal{N}]$ and suppose furthermore that the compact Lie group G acts on $\mathcal{M}$ leaving $\mathcal{N}$ invariant. Let $\mathcal{M}_n = \langle \mathcal{M}, e_1, e_2, \ldots, e_n \rangle''$ be the tower of factors obtained by iterating the Jones basic construction. From the fact that the Jones basic constructions are canonical it follows that the action of G extends to an action on $\cup_{n\geq 1} \mathcal{M}_n$ leaving each $\mathcal{M}_n$ invariant and in fact fixing each basic projection e_n. We first want to find conditions under which the basic tower for the inclusion of the fixed point algebras $\mathcal{N}^G \subset \mathcal{M}^G$ is obtained by taking fixed points of the tower for the inclusion $\mathcal{N} \subset \mathcal{M}$. In fact we have the following 'invariance principle':

Lemma **Suppose that $\mathcal{M} \rtimes G$ and $\mathcal{N} \rtimes G$ are both factors. Then $\mathcal{M}^G$ and $\mathcal{N}^G$ are both finite factors with $[\mathcal{M}^G:\mathcal{N}^G] = [\mathcal{M}:\mathcal{N}]$ and the basic tower for the inclusion $\mathcal{N}^G \subset \mathcal{M}^G$ is isomorphic to the one obtained by successively adjoining the basic projections e_1, e_2, ... to $\mathcal{M}^G$. The nth factor in the tower**

$$\langle \mathcal{M}^G, e_1, e_2, \ldots, e_n \rangle''$$

can be naturally identified with $\mathcal{M}_n^G$.

Proof Since the fixed point algebra is always a reduction of the crossed product, the first assertion is immediate. Since the crossed product $\mathcal{N} \rtimes G$ is a factor, we know by Connes' 2×2 matrix trick that $H^1(G, M_n(\mathcal{N}))$ is trivial and hence that every $\pi \in \hat{G}$ admits a unitary eigenmatrix $M(\pi)$ in $\mathcal{N} \otimes End(V_\pi)$. In view of the ampliation result of Pimsner and Popa and their result on the uniqueness of the basic construction, it suffices to prove the result for the first basic construction, since the general result follows by induction (or iteration).

We are thus reduced to proving that the factor $\mathcal{M}_1^G$ is generated as a von Neumann algebra by e_1 and $\mathcal{M}^G$. Now on the one hand we know that elements of the form ae_1b^* with a and b in $\mathcal{M}$ are total in $\mathcal{M}_1$. Thus if E denotes the conditional expectation onto the fixed point algebra of G obtained by averaging over G, it follows that elements of the form $E(ae_1b^*)$ are total in $\mathcal{M}_1^G$. On the other hand we know that elements of the form $xM_{ij}(\pi)$ with $x \in \mathcal{M}^G$ are total in $\mathcal{M}$, since the $M(\pi)$'s form a complete set of unitary eigenmatrices in $\mathcal{N}$ and hence $\mathcal{M}$. Therefore taking $a = xM_{ij}(\pi)$ and $b = yM_{rs}(\sigma)$ with $x, y \in \mathcal{M}^G$, we have

$$E(ae_1b^*) = E(xM_{ij}(\pi)e_1M_{rs}(\sigma)^*y^*) = xE(M_{ij}(\pi)M_{rs}(\sigma)^*)e_1y^*$$

which is patently in the von Neumann algebra generated by $\mathcal{M}^G$ and e_1. This completes the proof.

Let us now give three examples where the invariance principle may be applied to compute higher relative commutants.

Example 1. Suppose that $\alpha: G \to Aut(\mathcal{R})$ is a minimal action of G on $\mathcal{R}$ in the sense of Section IV. Let $\sigma: G \to U(W)$ be any finite dimensional (possibly reducible) unitary representation of G and define $\mathcal{N} = \mathcal{R}$ and $\mathcal{M} = \mathcal{R} \otimes End(W)$, endowed with the action $\alpha \otimes Ad\,\sigma$. We now observe that the canonical Jones tower for the inclusion of finite-dimensional factors $\mathbf{C} \subset End(W)$ is just

$$\mathbf{C} \subset End(W) \subset End(W)^{\otimes 2} \subset End(W)^{\otimes 3} \subset \cdots$$

with the basic projection given by the formula

$$e_n = (dim(W))^{-1} \sum_{ij} e_{ij}^n \otimes e_{ij}^{n+1}$$

where e_{ij}^m are matrix units in the mth factor. Note that if we make G act on the infinite tensor product $\otimes_{i=1}^{\infty} End(W)$ by the action $Ad\sigma \otimes Ad\overline{\sigma} \otimes Ad\sigma \otimes Ad\overline{\sigma} \otimes \cdots = \otimes Ad(\sigma \otimes \overline{\sigma})$, then the projections e_n are all fixed by G. Consequently, if we now tensor this tower by $\mathcal{R}$ with the obvious tensor product action, we obtain the tower and basic projections for the equivariant inclusion $\mathcal{R} \subset \mathcal{R} \otimes End(W)$. We may now apply the lemma. Thus the index of $\mathcal{N} = \mathcal{R}^{\alpha}$ in $\mathcal{M} = (\mathcal{R} \otimes End(W))^{\alpha \otimes Ad\sigma}$ is $dim(W)^2$ and the tower obtained by iterating the basic construction for the inclusion $\mathcal{R}^{\alpha} \subset (\mathcal{R} \otimes End(W))^{\alpha \otimes Ad\sigma}$ is given by

$$\begin{aligned}\mathcal{R}^{\alpha} &\subset (\mathcal{R} \otimes End(W))^{\alpha \otimes Ad\sigma} \\ &\subset (\mathcal{R} \otimes End(W) \otimes End(W))^{\alpha \otimes Ad\sigma \otimes Ad\overline{\sigma}} \\ &\subset (\mathcal{R} \otimes End(W) \otimes End(W) \otimes End(W))^{\alpha \otimes Ad\sigma \otimes Ad\overline{\sigma} \otimes Ad\sigma} \\ &\subset \cdots\end{aligned}$$

Bearing in mind that $\mathcal{R} \cap (\mathcal{R}^G)' = \mathbf{C}$, we see that the tower of relative commutants of the common subfactor $\mathcal{R}^{\alpha}$ is given by

$$\begin{aligned}End(W)^{Ad\sigma} &\subset (End(W) \otimes End(W))^{Ad\sigma \otimes Ad\overline{\sigma}} \\ &\subset (End(W) \otimes End(W) \otimes End(W))^{Ad\sigma \otimes Ad\overline{\sigma} \otimes Ad\sigma} \\ &\subset \cdots\end{aligned}$$

In particular $\mathcal{N}$ has trivial relative commutant in $\mathcal{M}$ if and only if W is an irreducible representation of G. The computation of the tower of relative commutants of $\mathcal{M}^G$ will be left as an exercise for the interested reader in this and the subsequent examples.

Example 2. Let us observe that it was unnecessary to assume that $\mathcal{M}$ and $\mathcal{N}$ were factors in the proof of the lemma, but merely that their crossed products by G were. So suppose now that $\alpha: G \to Aut(\mathcal{R})$ is an outer action of the finite group G and that H is a subgroup of G. Let $\mathcal{N} = \mathcal{R}$ and $\mathcal{M} = \mathcal{R} \otimes \ell^{\infty}(G/H)$ with the obvious diagonal action of G and the unique G-invariant trace. It is not hard to see that if X is a finite set, then the canonical Jones tower for the inclusion of finite dimensional algebras $\mathbf{C} \subset \ell^{\infty}(X)$ (with the canonical traces) is as follows

$$\mathbf{C} \subset \ell^{\infty} \subset End(\ell^2) \subset End(\ell^2) \otimes \ell^{\infty} \subset End(\ell^2)^{\otimes 2} \subset End(\ell^2)^{\otimes 2} \otimes \ell^{\infty} \subset \cdots$$

where $\ell^2 = \ell^2(X)$, $\ell^\infty = \ell^\infty(X)$ and the projections for the successive basic constructions are the obvious ones, namely alternately $\frac{1}{|X|}\sum e_{xy}$ and $\sum e_{xx} \otimes e_{xx}$. Of course when $X = G/H$ all these projections are G–invariant, and by the immediate extension of Jones' basic construction to inclusions of finite direct sums of type II_1 factors (again defined by inclusion matrices), one sees that the tower and basic projections for the inclusion of $\mathcal{R}$ in $\mathcal{R} \otimes \ell^\infty(X)$ are just obtained by taking the tensor product of the corresponding objects for the inclusion of $\mathbb{C}$ in $\ell^\infty(X)$. Now observe that if $\mathcal{M}$ is a G–algebra then $(\mathcal{M} \otimes \ell^\infty(G/H))^G$ admits a natural identification with $\mathcal{M}^H$ (compatible with the inclusion of $\mathcal{M}^G$). Now let G be a finite group with subgroup H and let $\alpha: G \to Aut(\mathcal{R})$ be an outer (so minimal) action of G. Then the index of $\mathcal{R}^G$ in R^H is $[G:H]$, the tower obtained by iterating the basic construction is given by

$$\begin{aligned}
\mathcal{R}^G &\subset \mathcal{R}^H \\
&\subset (\mathcal{R} \otimes End(\ell^2(G/H)))^G \\
&\subset (\mathcal{R} \otimes End(\ell^2(G/H)))^H \\
&\subset (\mathcal{R} \otimes End(\ell^2(G/H))^{\otimes 2})^G \\
&\subset (\mathcal{R} \otimes End(\ell^2(G/H))^{\otimes 2})^H \\
&\subset \cdots
\end{aligned}$$

and the tower of higher relative commutants is given by

$$\begin{aligned}
\mathbb{C} &\subset (End(\ell^2(G/H)))^G \\
&\subset (End(\ell^2(G/H)))^H \\
&\subset (End(\ell^2(G/H))^{\otimes 2})^G \\
&\subset (End(\ell^2(G/H))^{\otimes 2})^H \\
&\subset \cdots
\end{aligned}$$

This example was independently computed by Ocneanu using his bimodule picture of the higher relative commutants. He considered the equivalent inclusion of crossed products $\mathcal{N} = \mathcal{R} \rtimes H \subset \mathcal{M} = \mathcal{R} \rtimes G$ and observed that any choice of coset representatives for H in G provides an explicit Pimsner–Popa basis for $\mathcal{M}$ over $\mathcal{N}$. Thus the tensor products $\mathcal{M} \otimes_{\mathcal{N}} \mathcal{M} \otimes_{\mathcal{N}} \ldots \otimes_{\mathcal{N}} \mathcal{M}$ become easy to compute. On the other hand, our method can be extended to compact groups (by considering the inclusion $\mathbb{C} \subset L^\infty(G/H)$) where there is no really convenient basis available. We shall illustrate this in the next example by taking H to be the trivial subgroup.

Example 3. We shall now show how to compute the tower of higher relative commutants for an inclusion $\mathcal{M}^G \subset \mathcal{M}$ where G is a compact group with a minimal action on the factor $\mathcal{M}$. Indeed we may identify $\mathcal{M}$ with $(\mathcal{M} \otimes L^\infty(G))^G$ and reduce the computation to that of the (equivariant) inclusion $\mathbb{C} \subset L^\infty(G)$. The tower obtained by iterating the basic construction here is the exact analogue of what we got in the case that G was finite. Thus we obtain

$$\mathbb{C} \subset L^\infty(G) \subset \mathcal{B}(L^2(G)) \subset \mathcal{B}(L^2(G)) \otimes L^\infty(G) \subset \mathcal{B}(L^2(G)) \otimes \mathcal{B}(L^2(G)) \subset \cdots$$

It follows that the tower for the iterated basic constructions for the original inclusion $\mathcal{M}^G \subset \mathcal{M}$ is just

$$\mathcal{M}^G \subset \mathcal{M} \subset \mathcal{M} \rtimes G \subset \mathcal{M} \otimes \mathcal{B}(L^2(G)) \subset \cdots,$$

and from this it is easy to work out the higher relative commutants. One finds that the higher relative commutants of $\mathcal{M}^G$ are stably isomorphic to either $\mathcal{R}(G)$ or $\mathbb{C}$; while those of $\mathcal{M}$ are stably isomorphic to either $L^\infty(G)$ or $\mathbb{C}$.

VII. Yang–Baxter Representations of $S(\infty)$ and Ergodic Actions

In this final section, we shall explain in a rather vague and speculative way how our theory of bicharacters runs parallel to the theory of Quantum Groups, in the sense that it may be regarded as a bounded version of Drinfeld's theory. We will indicate in particular how the quantum Yang–Baxter equation for bicharacters — which seemed almost to arise almost by accident — in fact plays a crucial rôle, in that any 'local' solution of this equation automatically gives rise to a bicharacter. Using the fact that any homomorphism of compact groups $\theta: G_1 \to G_2$ gives rise to a functorial map $\theta^*: H^2(\hat{G}_1, \mathbb{T}) \to H^2(\hat{G}_2, \mathbb{T})$, we lose no generality by restricting our discussion to the case when G is the unitary group $U(n)$ or even just $SU(n)$ or $PU(n)$. We find that a bicharacter $\beta \in \mathcal{R}$ is uniquely determined by some (π, π)–component $B = \beta(\pi, \pi) \in End(V_\pi \otimes V_\pi)$ if π is a faithful irreducible representation of G, where B just satisfies the quantum Yang–Baxter equation and certain extra unitarity conditions. In principle it is possible to use the method of Drinfeld, axiomatised by Lyubashenko, to recover the whole bicharacter from the matrix B. This approach is very close in spirit to the strategy of S. Doplicher and J. Roberts for recovering a compact group from its associated monoidal C* category. It should also be added that in their previous work on quantum field theory, these authors had already noticed the appearance of a version of the QYBE as a type of obstruction in their theory and that it assumed a formal similarity with the classification of ergodic actions of compact groups. In their approach, however, the equation assumed no physical significance, so was not investigated further. We propose an alternative method of recovering the bicharacter from the local solution of the QYBE. We shall associate with any such solution a finite 'Yang–Baxter' factor representation π of the infinite symmetric group S_∞ in the hyperfinite II_1 factor $\mathcal{R}$. We then consider the subfactor $\mathcal{N} = \pi(S_\infty)''$ of $\mathcal{M} = \mathcal{R}$. The existence of such a representation of S_∞ had already been noted by the Russian school, although they only considered its restriction to the finite subgroups S_n rather than looking at the 'asymptotic' properties of the whole representation π. It turns out that $\mathcal{N}$ has trivial relative commutant in $\mathcal{M}$ and has infinite index. It seems very likely that the corresponding relative commutant towers have depth two and that for example $\mathcal{N}' \cap \mathcal{M}_1$ is isomorphic to $\mathcal{R}(\mathcal{PU}(V_\pi))$. On the other hand Ocneanu's 'paragroup' structure on the relative commutant towers should allow one to recover the Kac algebra structure on this algebra and its dual. For example it is already clear that one has a tensor product structure on the irreducible representations of $\mathcal{N}' \cap \mathcal{M}_1$ — it is just the usual tensor product on $\mathcal{N}$–bimodules. One would also expect the algebra $A_\omega(G)$ to appear in the picture. What is not so clear at present is whether the homomorphism Λ or for that matter the algebra $\pi_\omega(\hat{G})''$ will figure.

Before entering into the realm of wild speculation, let us now explain what we can actually do at present. Let $\beta \in \mathcal{R} \otimes \mathcal{R}$ be a bicharacter of $\hat{G}$ corresponding to some cocycle ω of $\hat{G}$. We fix $\pi \in \hat{G}$ and set $V = V_\pi$. Let $B \in End(V \otimes V)$ be the (π, π)–component of β. Then the Yang–Baxter equation for B is just $B_{12}B_{13}B_{23} = B_{23}B_{13}B_{12}$. Now we note that by definition $\beta = \sigma\omega^*\omega$. Let $\Omega \in End(V \otimes V)$ be the (π, π)–component of ω and let $S \in End(V \otimes V)$ be the usual flip, $S(v_1 \otimes v_2) = v_2 \otimes v_1$. Evidently $B = S\Omega^*S\Omega$. As is customary, we next introduce the associated R–matrix $R = SB = \Omega^*S\Omega$. Since it is conjugate to the flip, R is a unitary involution. Moreover the Yang–Baxter equation takes a particularly simple form in terms of the R–matrix: $R_{12}R_{23}R_{12} = R_{23}R_{12}R_{23}$. Since R has square one, this condition simply says that $(R_{12}R_{23})^3 = I$. This shows that the matrices R_{12} and R_{23} satisfy the standard relations for the generators of the symmetric group S_3, with R_{12} and R_{23} corresponding to the permutations $(1,2)$ and $(2,3)$ respectively. Thus we get a representation of S_3 in $End(V) \otimes End(V)$. Now let A be the infinite tensor product $\otimes_{n=1}^{\infty} End(V)$. We can define a representation π_R of the infinite symmetric group in A by making the R–matrix 'propagate'. Indeed if we define $\pi_R((n, n+1)) = R_{n,n+1}$ for $n \geq 1$, then this extends uniquely to a representation of S_∞ since the matrices $\sigma_n = R_{n,n+1}$ satisfy the defining relations for S_∞, namely $\sigma_n^2 = 1 = (\sigma_n\sigma_{n+1})^3$ for all n and $\sigma_n\sigma_m = \sigma_m\sigma_n$ whenever $|n - m| > 1$. Such representations of S_∞ will be called *Yang–Baxter representations.* If we take the natural trace τ on A then it τ restricts to a trace on $\pi(S_\infty)$. It follows that if γ_1 and γ_2 are disjoint permute in S_∞, then $\tau(\pi_R(\gamma_1\gamma_2)) = \tau(\pi_R(\gamma_1))\tau(\pi_R(\gamma_2))$. To see this we note that we may conjugate the permutations so that γ_1 only moves $1, 2, \ldots, N$ and γ_2 fixes this set. The traces do not change when we do this and the multiplicativity becomes manifest. It follows that τ defines an extremal trace of S_∞ so that $\mathcal{N}_R = \pi_R(S_\infty)''$ is a subfactor of $\mathcal{M} = \pi_\tau(A)''$. Let us now show that π_R and π_S are quasi–equivalent representations of S_∞, that is the natural map $\pi_R(\sigma) \mapsto \pi_S(\sigma)$ extends (uniquely) to an isomorphism of $\mathcal{N}_R$ onto $\mathcal{N}_S$. For this we only have to check that the corresponding traces agree. By multiplicativity this reduces to showing that $\tau(\pi_R((1, \ldots, n))) = (dim(V))^{-n+1}$ for every $n \geq 2$. Now the conditions $\alpha \otimes \iota(\beta) = \beta^*$ and $\delta_G(e_1)\beta = \delta_G(e_1)$ imply that the matrix R satisfies $Tr \otimes \iota(R) = I = \iota \otimes Tr(R)$. In other words

$$\sum_i R_{ii,ab} = \delta_{ab} = \sum_i R_{ab,ii}. \qquad (*)$$

(Here if e_i is a basis in V, we define the matrix $R_{ab,cd}$ of R by the formula $R(e_i \otimes e_j) =$

$\sum_{k\ell} R_{ki,\ell j} e_k \otimes e_\ell$.) On the other hand (using the summation convention)

$$\begin{aligned}\tau(\pi_R((1,2,\ldots,n))) &= \tau(R_{12}R_{23}\cdots R_{n-1,n})\\ &= (dim(V))^{-n+1} R_{ii,ab}R_{ba,cd}R_{dc,ef}\cdots R_{yx,jj}\\ &= (dim(V))^{-n+1},\end{aligned}$$

by repeated application of the identity $(*)$. Thus as far as representations go π_R and π_S are indistinguishable; however, if the isomorphism of $\mathcal{N}_R$ onto $\mathcal{N}_S$ extends to an automorphism of $\mathcal{M}$ onto $\mathcal{M}$ then it is not too hard to see that R has to have the form $(v\otimes v)S(v^*\otimes v^*)$ for some unitary v of V. Thus R will be trivial. It is therefore of paramount importance to examine the precise position of $\mathcal{N}_R$ in $\mathcal{M}$, i.e. to work out the towers of higher relative commutants. We would like as far as possible to immitate the methods used for π_S in the previous sections. For the higher relative commutants this could be facilitated by exhibiting a dual action of the Kac algebra $\mathcal{G} = (\mathcal{R}(\mathcal{PU}(V)), \delta_\omega)$ on $\mathcal{M}$ with $\mathcal{N}_R$ as fixed point algebra. We would therefore also need to find an equivariant copy of the dual $\hat{\mathcal{G}} = A_\omega(\mathcal{PU}(V))$ in $\mathcal{M}$ and this would need some extension of Weyl duality for the actions on $V^{\otimes n}$ of S_n (via π_R) and $\mathcal{G}$ (via the tensor product induced by δ_ω). Using Connes' 2×2 matrix trick for producing eigenmatrices together with Ocneanu's announced result on the vanishing of 2–cohomology for centrally free cocycle actions of discrete Kac algebras, we have at least some moral foundation for this copy of $\hat{\mathcal{G}}$ to exist. We could then try to construct the Jones tower for the inclusion $\mathcal{N}_R \subset \mathcal{M}$ by an analogue of the invariance principle of the last section. It would be equally interesting to do the analogous computations for non–symmetric solutions of the QYBE using some version of the invariance principle. These subfactors correspond to factor representations of the infinite braid group or equivalently the infinite Hecke algebra $H_\infty(q)$ and the trace has in general to be replaced by a Powers state to ensure unitarity. Consequently the machinery of Kosaki would have to be used. Unfortunately although this method would seem likely to work for the Pimsner–Popa subfactors (when q is real), the really interesting values — namely when q is a root of unity — would apparently remain inaccessible by this method. Using a quite different method based on a priori bounds on the dimensions of the higher relative commutants, however, Hans Wenzl has succeeded in computing the higher relative commutant towers even in these exceptional cases. It remains a tantalising task to obtain his results by some form of the invariance principle.

At present we have only checked that the relative commutant $\mathcal{N}_R' \cap \mathcal{M}$ is trivial. In order to explain this computation, we start by observing that the unitarity properties of $\beta(\pi,\overline{\pi})$ lead to the following new identities for the R–matrix:

$$\sum_{ij} R_{ij,ab}R_{ji,cd} = \delta_{ad}\delta_{bc} = \sum_{ij} R_{ab,ij}R_{cd,ji}. \qquad (**)$$

Exactly as in Section V, we will show that the relative commutant is trivial by proving that the conditional expectation onto it is scalar–valued. This we accomplish essentially by exhibiting the conditional expectation as a weak limit of averages over the finite symmetric groups, although we shall need an extra 'doubling' trick to make the computation efficiently. Let us start with a simple case by showing that if x is an arbitrary element of finite support in A and y is an element in the first copy of $End(V)$, then $\lim_{n\to\infty} \tau(xAv_{S_n}(y)) = \tau(x)\tau(y)$. This will show that the conditional expectation $E_{\mathcal{N}_R'\cap\mathcal{M}}(y)$ of y is just the scalar $\tau(y)$. Suppose then that the element x lies in the first m copies of $End(V)$. Since the subgroup of permutations fixing 1 also commutes with y in the representation π_R, we may replace Av_{S_n} by a sum over convenient coset representatives of this subgroup. Let $\gamma_t = (1,2,\ldots,t) = (1,2)(2,3)\cdots(t-1,t)$ so that γ_t^{-1} $(1 \le t \le n)$ constitutes a complete set of coset representatives. We therefore have

$$\begin{aligned}\lim_{n\to\infty} \tau(xAv_{S_n}(y)) &= \lim_{n\to\infty} \frac{1}{n}\sum_{t=1}^{n} \tau(x\pi_R(\gamma_t^{-1})y\pi_R(\gamma_t))\\ &= \lim_{n\to\infty} \frac{1}{n}\sum_{t=1}^{n} \tau(x\pi_R((t-1,t)\cdots(12))y\pi_R((1,2)\cdots(t-1,t)))\\ &= \lim_{n\to\infty} \frac{1}{n}\sum_{t=m+1}^{n} \tau(x\pi_R((t-1,t)\cdots(12))y\pi_R((1,2)\cdots(t-1,t)))\\ &= \tau(x\pi_R(\gamma_{m+1}^{-1})y\pi_R(\gamma_{m+1}))\\ &= \tau(x(R_{m,m+1}\ldots R_{12})y(R_{12}\ldots R_{m,m+1})).\end{aligned}$$

On the other hand, m successive applications of formula (**) show that this last expression is equal to $\tau(x)\tau(y)$, as claimed.

As soon as we try to extend the above reasoning to cover elements x and y with arbitrary finite supports, we encounter a difficulty characteristic of the Yang–Baxter representation (or the analogous representations of Wenzl of the Hecke algebras of Type A_n generated by the Jones basic projections e_i). The operators $\pi_R((i,j))$ with i and j far apart are not local, i.e. they are not given by the naive expressions $R_{i,j}$. Instead they must be expressed in terms of the 'simple' transpositions $(i,i+1)$. This is because the natural *order* on the positive integers is crucial in the definition of π_R, whereas it was irrelevant for π_S. The reader will recall that exactly the same problem arose for V. Jones in writing down an explicit basis for the Hecke algebras. To overcome this difficulty we shall make use of a simplified version of the Murakami–Wenzl 'cabling' trick, originally used to facilitate the computation of higher link invariants from solutions of the QYBE. In our context this means that for fixed k, we regard the set of positive integers as being a disjoint union of k copies of itself in a manner consistent with the natural order. Thus we have the k identification maps $\theta_j(n) = k(n-1)+j$ where $1 \le j \le k$, which have the effect of reconstructing the set of positive integers by 'interleaving' k copies of itself. Clearly $\theta_n \cdot \theta_m = \theta_{nm}$. Moreover, each map θ_n induces a homomorphism θ_n^* of S_∞ into itself. Thus for $n = 2$ the transposition $(i,i+1)$ is mapped to $(2i-1,2i+1)(2i,2i+2)$ under θ_2^*, so that θ_2^* just amounts to a 'doubling' of the permutation representation; and if we double twice, we just quadruple the representation, and so on. It will be important to know what happens to the Yang–Baxter representation if we compose it with the doubling operation. In fact the expression $(13)(24) = (23)(12)(34)(23)$ immediately shows that $\pi_R \cdot \theta_2^* = \pi_{R^{(2)}}$, where $R^{(2)}$ is the R–matrix $R_{23}R_{12}R_{34}R_{23}$ on $(V \otimes V) \otimes (V \otimes V)$. That the doubled R–matrix satisfies the same conditions $(*)$ and $(**)$ as the original matrix follows by noting that the corresponding matrix B is just the $(\pi \otimes \pi, \pi \otimes \pi)$–component of $\delta_\omega \otimes \delta_\omega(\beta)$ (in an obvious sense). Furthermore if we successively double the R–matrix, these conditions are still satisfied.

Now what we have shown so far is that if y comes from the first copy of $End(V)$ in A, then $E_{\mathcal{N}_R' \cap \mathcal{M}}(y) = \tau(y)$. Suppose now that y has finite support, so lies in the first 2^k copies of $End(V)$ in A. Then y lies in the first copy of $(End(V))^{\otimes 2^k} \equiv End(V^{\otimes 2^k})$ in A. So applying our simple result to the k–fold doubling $R' = R^{(2^k)}$ of R, we find that $E_{\mathcal{N}_{R'}' \cap \mathcal{M}}(y) = \tau(y)$. Since by definition $\pi_{R'}(S_\infty) \subseteq \pi_R(S_\infty)$, we see that $\mathcal{N}_{R'}' \cap \mathcal{M} \supseteq \mathcal{N}_R' \cap \mathcal{M}$; since conditional expectations are functorial with respect to inclusions, we deduce that $E_{\mathcal{N}_R' \cap \mathcal{M}}(y) = \tau(y)$. By continuity the same holds true for any $y \in \mathcal{M}$, and so $\mathcal{N}_R$ does indeed have trivial relative commutant.

We conclude with some general remarks. Let G be a compact Abelian group. Now the exponential maps $\mathbb{C} \to \mathbb{C}^*$ and $\mathbb{R} \to \mathbb{T}$ induce natural maps $H^2(\hat{G}, \mathbb{C}) \to H^2(\hat{G}, \mathbb{C}^*)$ and $H^2(\hat{G}, \mathbb{R}) \to H^2(\hat{G}, \mathbb{T})$ which we shall view as providing 'quantisation' maps from an additive theory to a multiplicative theory. This terminology is not as fanciful as it might appear, since as we have seen the multiplicative theory essentially measures the 'global' deformations of the Hopf algebras $\mathcal{U}(G)$ and $\mathcal{R}(G)$ (or their duals), while the additive theory is precisely the infinitesimal linearised version of these deformations. In other words it is defined by the equations of the tangent space at the identity of the multiplicative theory. These defining equations can be obtained by differentiating the cocycle relations for a one–parameter family of cocycles emanating from the trivial identity cocycle. All this makes sense if one replaces G by a non–commutative compact group and takes the appropriate Hopf algebras for the unbounded or bounded theory. In the case of the classical groups, the unbounded theory amounts precisely to Drinfeld's theory of quantum groups. The corresponding linearised theory is just the classical Yang–Baxter equation. There remains the problem of quantising the classical solutions to obtain the quantum solutions: Drinfeld has shown that there is no (formal) obstruction to doing this. It should be apparent that the bounded unitary theory that we have developed shares many similarities with the theory of Drinfeld and his school; in particular our theory has contact with the work of Gurevich and Lyubashenko on the symmetric QYBE. Our approach — through ergodic actions and subfactors — might possibly lead to some new insights. For example the so–called finite groups of central type, namely those admitting full multiplicity ergodic actions on matrix algebras, provide new solutions of the symmetric QYBE. (These groups are not particularly well understood, although they are known to be solvable, by the classification of finite simple groups.) Furthermore the methods of Section V might possibly extend to higher rank classical groups to give a classification of the solutions of the symmetric QYBE. The solutions we are considering are similar to the flip and therefore, as Gurevich points out, have the right signature to be interesting from the point of view of supersymmetry.

One fairly plausible conjecture about cocycles and bicharacters (or the corresponding full multiplicity ergodic actions) is that if they are sufficiently near to the trivial cocycle, then they should be induced from a maximal torus. As support for this let us verify this conjecture on the infinitesimal level, using an argument presumably familiar to Bellavin and Sklyanin but unrecorded in the literature. The classical version of the QYBE is just obtained by taking a C^1 family $B(t)$ such that $B_{12}(t)B_{13}(t)B_{23}(t) = B_{23}(t)B_{13}(t)B_{12}(t)$ for all t and $B(0) = I$ and defining $b \in End(V \otimes V)$ by $b = \dot{B}(0)$. Thus b is skew adjoint, $b_{12} = -b_{21}$ and b satisfies the classical Yang–Baxter equation: $[b_{12}, b_{13}] + [b_{12}, b_{23}] + [b_{13}, b_{23}] = 0$. Now we define a map $r: End(V) \to End(V)$ via $Tr(r(X)Y) = Tr(b(X \otimes Y))$. Actually we shall only consider the restriction of r to the Lie algebra $\mathcal{L}$ of skew–adjoint matrices with its tracial inner product. r is then a skew–symmetric endomorphism of $\mathcal{L}$ such that $[r(X), r(Y)] = r([r(X), Y] + [X, r(Y)])$ for all $X, Y \in \mathcal{L}$. Thus if r were invertible, its inverse would define a derivation of $\mathcal{L}$. On the other hand, the only compact Lie algebras with invertible derivations are Abelian. So we end by observing that the image of r is a Lie subalgebra $\mathcal{L}_0$ of $\mathcal{L}$. Moreover the inverse of the restriction of r to $\mathcal{L}_0$ defines an invertible derivation of $\mathcal{L}_0$, so that $\mathcal{L}_0$ must therefore be Abelian. It is then easy to see that b is induced from this subalgebra (in an obvious sense) and hence from a maximal torus, just as we claimed.

References

1. V. G. Drinfeld, *Quantum Groups*, Seminar on Supermanifolds **4**, ed. D. Leites, Stockholm, 1987.
2. D. J. Gurevich, *Quantum Yang–Baxter equation and a generalization of the formal Lie theory*, Seminar on Supermanifolds **4**, ed. D. Leites, Stockholm, 1987.
3. L. Hodgkin, *The equivariant Künneth theorem in K–theory*, in "Topics in K–theory," Lect. Notes in Math., **496**, Springer–Verlag, 1975.
4. R. Høegh–Krohn, M. Landstad and E. Størmer, *Compact ergodic groups of automorphisms*, Ann. Math. **114** (1981), 75–86.
4. V. F. R. Jones, *Index for subfactors*, Invent. Math. **72** (1983), 1–25.
5. G. G. Kasparov, *K–functor and extensions of C* algebras*, Iz. Akad. Nauk. SSSR. **44** (1980),571–636.
6. G. G. Kasparov, *K–theory, group C* algebras and higher signatures*, Conspectus, Chernogolovka, 1983.
7. V. V. Lyubashenko, *Vectorsymmetries*, Seminar on Supermanifolds **14**, ed. D. Leites, Stockholm, 1987.
8. A. Ocneanu, "Actions of Discrete Amenable Groups on von Neumann algebras," Lect. Notes in Math., **1138**, Springer–Verlag, 1985.
9. A. Ocneanu, *Classification of subfactors of finite index and finite depth*, in preparation.

10. M. Pimsner and S. Popa, *Entropy and index for subfactors*, Ann. Sci. Ec. Norm. Sup. **19** (1986), 57–106.

11. M. Pimsner and S. Popa, *Iterating the basic construction*, preprint, Increst, 1985.

12. H. Upmeier, *Toeplitz C^* algebras on bounded symmetric domains*, Ann. of Math. **119** (1984), 549–576.

13. H. Upmeier, *Index theory for Toeplitz operators on bounded symmetric domains*, Bull. A.M.S. **16** (1987), 109–112.

14. A. J. Wassermann, "Automorphic Actions of Compact Groups on Operator Algebras," Ph. D. Dissertation, University of Pennsylvania, 1981.

15. A. J. Wassermann, *Algèbres d'opérateurs de Toeplitz sur les groupes unitaires*, C. R. Acad. Sc. Paris **299** (1984), 871–874.

16. A. J. Wassermann, *Ergodic Actions of Compact Groups on Operator Algebras: I: General Theory*, submitted for publication; *II: Classification of Full Multiplicity Ergodic Actions*, to appear in Can. J. Math.; *III: Classification for $SU(2)$*, to appear in Invent. Math.

17. A. J. Wassermann, *Equivariant K–theory II: Hodgkin's spectral sequence in Kasparov's bivariant theory*, preprint, Liverpool University, 1987.

18. A. J. Wassermann, *Product type actions of compact Lie groups I, II*, to appear in J. Operator Theory.

Derived link invariants and subfactors

Hans Wenzl
University of California, Berkeley

We construct from given representations of braid groups new representations. This can be used to construct new link invariants, new subfactors and new $\check{R}$ matrices from given subfactors, link invariants and $\check{R}$ matrices respectively.

As usual, let B_∞ be the infinite braid group, given by generators $\sigma_1, \sigma_2, \ldots$ and relations

(B1) $\quad \sigma_i\sigma_{i+1}\sigma_i = \sigma_{i+1}\sigma_i\sigma_{i+1}$ for $i = 1, 2, \ldots$ and
(B2) $\quad \sigma_i\sigma_j = \sigma_j\sigma_i \quad$ if $|i-j| \geq 2$.

Furthermore we denote the subgroup of B_∞, which is generated by σ_r, $\sigma_{r+1} \ldots \sigma_{s-2}$, σ_{s-1} by $B_{r,s}$. We will just write B_f for $B_{1,f}$. Let now the braid $\sigma_i^{(f)}$ be given by the picture (see (Birman) for details)

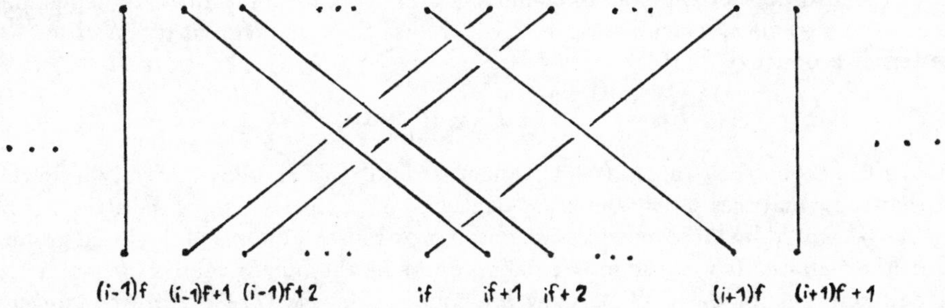

$(i-1)f \quad (i-1)f+1 \quad (i-1)f+2 \quad if \quad if+1 \quad if+2 \quad (i+1)f \quad (i+1)f+1$

We note here that our theory also works if we vary $\sigma_i^{(f)}$ by either conjugating it by other braids or by multiplying it by central elements of $B_{(i-1)f+1,if}$ or $B_{if+1,(i+1)f}$ (see for instance (Murakami)). The results, however, will be essentially the same as for our special choice.

It is easy to check by pictures that $\sigma_i \mapsto \sigma_i^{(f)}$ induces an injective homomorphism from B_∞ into itsself. If ρ is an arbitrary representation of B_∞, we obtain a new representation $\rho^{(f)}$ by

$$\rho^{(f)}(\sigma_i) = \rho(\sigma_i^{(f)}).$$

In the sequel, we always assume that ρ is an approximately finite dimensional (= AFD) representation of B_∞, i.e. $\rho(CB_n)$ is finite dimensional for all $n \in \mathbf{N}$ and we also assume that these finite dimensional representations are semisimple.

We are now going to decompose these representations. Let p be a minimal idempotent in $\rho(\mathbf{C}B_f)$. Observe that conjugation by $\sigma_i^{(f)}$ interchanges σ_j and σ_{j+f} for $j = (i-1)f+1,\ (i-1)f+2,\ \ldots,\ if-1$. We use this to define inductively

$p_1 = p$ and
$p_i = (\rho^{(f)}(\sigma_{i-1}))p_{i-1}(\rho^{(f)}(\sigma_{i-1}))^{-1}$.

Furthermore let $p^{(n)} = p_1 p_2 \ldots p_n$. Then we have

(a) $p_i \in \rho(\mathbf{C}B_{(i-1)f+1,if})$,
(b) $p_i p_j = p_j p_i$ for all $i, j \in \mathbf{N}$ and
(c) $\rho^{(f)}(B_n)$ commutes with $p^{(m)}$ for all $m \geq n$.

We will moreover assume that

$$(*) \qquad p^{(m)}\rho(\mathbf{C}B_{nf})p^{(m)} \cong p^{(n)}\rho(\mathbf{C}B_{nf})p^{(n)} \quad \text{for all } m > n.$$

We obtain from this injections

$$\rho^{(f)}(\mathbf{C}B_{nf})p^{(n)} \to \rho^{(f)}(\mathbf{C}B_{(n+1)f})p^{(n+1)} \to \rho^{(f)}(\mathbf{C}B_{(n+2)f})p^{(n+2)} \to \ldots$$

Taking the inductive limit, we obtain an AFD representation $\rho^{(f,p)}$ of B_∞.

As an application of these techniques, we can construct new solutions of the quantum Yang-Baxter equations (QYBE) from existing ones. We start with a classical Lie algebra acting on a vector space V in its standard representation. Using their quantization (following work of Kulish, Reshetikhin and Sklyanin) M. Jimbo obtained matrices $\check{R}$, depending on 2 parameters q and x, which act on $V \otimes V$. We use them to define matrices R_i on an infinite tensor product $\bigotimes^\infty V$ by

$$R_i \ = \ \ldots 1 \otimes \check{R}(q,0) \otimes 1 \ldots,$$

where $\check{R}(q,0)$ acts on the i-th and $(i+1)$-th factor of $\bigotimes^\infty V$. It follows directly from the QYBE that these matrices satisfy the braid relations. As soon as we know the structure of the algebra generated by these matrices, we obtain new representations of the braid groups by the method above. If we take in our definition of R_i the full $\check{R}$ matrix, we similarly obtain new solutions of the QYBE. They are solutions of the QYBE corresponding to higher representations of the given Lie algebra. In case of the classical Lie algebras of type A, it has already been observed by Jimbo that the algebras generated by the R_i's are quotients of the Hecke algebras of type A. For types B, C and D, one obtains quotients of a new algebra discovered in connection with Kauffman's link invariant for type B, C and D (see (Birman&Wenzl) and (Turaev)).

The main ingredient for getting a link invariant via the approach of Jones is a *Markov trace*, i.e. a trace tr on $\mathbf{C}B_\infty$ such that for all $n \in \mathbf{N}$ and $\beta \in B_n$ we have

$$tr(\sigma_n^{\pm 1}\beta) = tr(\sigma_n)tr(\beta).$$

Recall that any such trace defines a link invariant L_{tr} via Markov's theorem by

$$L_{tr}(\hat{\beta}) = tr(\sigma_1)^{1-n}tr(\beta),$$

where $\beta \in B_n$ and $\hat{\beta}$ is its closure (see (Birman)). We remark here that for representations of braid groups as $R\check{}$ matrices, one can define Markov traces easily by a deformation of the usual trace on $\bigotimes^\infty Gl(V)$ by a special density matrix. This was first observed in this context in (Pimsner&Popa) for the Jones polynomial, for which the matrices come from the deformation of sl_2 in standard form. It was only noticed after generalizing Pimsner and Popa's matrices to the sl_k case (which was also done independently by Pimsner and Popa) that their matrices are the same as the ones of Jimbo.

Let now tr be a Markov trace factoring over ρ. Under mild assumptions on tr (essentially just faithfulness) the condition $(*)$ for the construction of $\rho^{(f,p)}$ holds. Let $tr^{(p)}$ be the normalized trace which is obtained from tr by restriction to $\rho^{(f,p)}(B_n)$ and renormalization. It is easy to see that this restriction is well-defined independently of n.

To get a link invariant we only need to prove that $tr^{(p)}$ induces a Markov trace on $\mathbf{C}B_\infty$ via $\rho^{(f,p)}$. This can be shown for the Hecke algebra representations of the braid groups in the following way:

If we set the parameter q of the Hecke algebras equal to $e^{2\pi i/l}$, tr factors over special unitary representations $\pi^{(k,l)}$ of B_∞ for $k = 1, 2, \ldots\, l$ (see (Wenzl)). In this case, the Markov property of $tr^{(p)}$ can be shown easily using detailed results about the centralizer of the subfactor $\pi^{(k,l)}(\langle\, \sigma_{2f+1}, \sigma_{2f+2} \ldots \rangle)''$ of $\pi^{(k,l)}(\mathbf{C}B_\infty)''$. The general case follows from this immediately using elementary properties of analytic functions.

Theorem

There exists for each Young diagram λ with f boxes an AFD representation $\rho^{(\lambda)}$ of B_∞ which depends on a parameter q. Moreover, there exists a Markov trace $tr^{(\lambda)}$ depending on q and an additional parameter l which factors over $\rho^{(\lambda)}$ and which is faithful on its image for almost all values of the parameters (i.e. except for a set of measure 0). In particular, we obtain for each Young diagram λ as above a 2-variable link invariant.

In case of the Young diagram with one box, the corresponding representation is the Hecke algebra representation of the braid group and the link invariant is the 2-variable polynomial of (Freyd et al.).

Remarks:

1. If q is not a root of unity, the structure of the images of $\mathbf{C}B_n$ under these representations can be determined using the Littlewood-Richardson rule. This provides a method of proving that our newly constructed traces are distinct from the original one.

2. Similar results have been obtained independently by Murakami for another choice of $\sigma_i^{(f)}$.

3. As in the original case, we obtain for special values of the parameters unitary repesentations of B_∞ and examples of subfactors of the hyperfinite II_1 factor.

4. V. Jones independently defined new link invariants using Jimbo's solutions of the $QYBE$ corresponding to higher representations of sl_2. They can be deduced as special cases from the link invariants in our theorem similarly as his one variable polynomial can be deduced from the 2-variable polynomial in (Freyd et al.). Moreover, all link polynomials coming from solutions of $QYBE$ corresponding to any finite dimensional irreducible representation of sl_k for any $k \geq 2$ can be obtained as special cases of the invariants in our

theorem.

5. Similar results can also be obtained for the Kauffman polynomial using the algebras in (Birman&Wenzl) (see also (Murakami)). More precisely, we can again define for each Young diagram a 2-variable link invariant where the invariant corresponding to the diagram with one box is the Kauffman polynomial. As a consequence of Turaev's result, all link invariants which can be derived from solutions of the QYBE corresponding to finite dimensional representations of orthogonal or symplectic groups can be obtained from specializations of these polynomials.

References

Birman, J., Braids, links and mapping class groups, Ann. Math. Studies 82, Princeton Univ. Press, 1974.

Birman, J., Wenzl, H., Braids, link polynomials and a new algebra, preprint, Columbia University.

Freyd, P., Yetter, D., Hoste, J., Lickorish, W.B.R., Millett, K., Ocneanu, A., A new polynomial invariant of knots and links, Bull. AMS 12, No.2, April 1985, p. 239-246.

Jimbo, M., Quantum R-matrix for the generalized Toda system, Com. Math. Phys. 102,4 (1986), 537-547.

Jones, V.F.R., A polynomial invariant for knots via von Neumann algebras, Bull. AMS 12, No. 1, Jan. 1985, p. 103-111.

Kulish, P.P., Sklyanin, E.P., On the solutions of the Yang-Baxter equation, J. Soviet Math. 19 (1982).

Murakami, J., The r-parallel version of link invariants, preprint.

Pimsner, M., Popa, S., Entropy and index for subfactors, Ann. scient. Ec. Norm. Sup. 4^e serie, t. 19, (1986), p. 57-106.

Turaev, V., The Yang-Baxter equation and invariants of links, preprint, Steklov Institute.

Wenzl, H., Representations of Hecke algebras and subfactors, thesis, University of Pennsylvania (1985).